The Web of Life

Also by Fritjof Capra

The Tao of Physics
The Turning Point
Uncommon Wisdom
Belonging to the Universe

A New Scientific Understanding of Living Systems

Fritjof Capra

THE WEB OF LIFE

ANCHOR BOOKS
A DIVISION OF RANDOM HOUSE, INC.
New York

First Anchor Books Trade Paperback Edition, October 1997

 Published in the United States by Anchor Books, a division of Random House, Inc., New York, and simultaneously in Canada by Random House of Canada Limited, Toronto. Originally published in hardcover in the United States by Anchor Books, a division of Random House, Inc., in 1996.

The Library of Congress has cataloged the Anchor hardcover edition as follows:

Capra, Fritjof.
The web of life: a new scientific understanding of living systems
Fritjof Capra.—1st Anchor Books ed.
p. cm.
Includes bibliographical references and index.
1. Life (Biology) 2. Biological systems. 3. System theory. I. Title.
QH501.C375 1996
574'.01—dc20 96-12576
CIP

ISBN 0-385-47676-0

www.anchorbooks.com

Printed in the United States of America
20 19 18 17 16 15 14

Contents

To the memory of my mother,
Ingeborg Teuffenbach,
who gave me the gift and the discipline of writing.

This we know.
All things are connected
like the blood
which unites one family. . . .

Whatever befalls the earth,
befalls the sons and daughters of the earth.
Man did not weave the web of life;
he is merely a strand in it.
Whatever he does to the web,
he does to himself.

—TED PERRY, inspired by Chief Seattle

Acknowledgments

The synthesis of concepts and ideas presented in this book took over ten years to mature. During this time I was fortunate to be able to discuss most of the underlying scientific models and theories with their authors and with other scientists working in those fields. I am especially grateful

- to Ilya Prigogine for two inspiring conversations during the early 1980s about his theory of dissipative structures;
- to Francisco Varela for explaining to me the Santiago theory of autopoiesis and cognition in several hours of intensive discussions during a skiing retreat in Switzerland, and for numerous enlightening conversations over the past ten years about cognitive science and its applications;
- to Humberto Maturana for two stimulating conversations in the mid-1980s about cognition and consciousness;
- to Ralph Abraham for clarifying numerous questions regarding the new mathematics of complexity;
- to Lynn Margulis for an inspiring dialogue in 1987 about the Gaia hypothesis, and for encouraging me to publish my synthesis, which was then just emerging;

- to James Lovelock for a recent enriching discussion of a wide range of scientific ideas;
- to Heinz von Foerster for several illuminating conversations about the history of cybernetics and the origins of the concept of self-organization;
- to Candace Pert for many stimulating discussions of her peptide research;
- to Arne Naess, George Sessions, Warwick Fox, and Harold Glasser for inspiring philosophical discussions; and to Douglas Tompkins for urging me to go deeper into deep ecology;
- to Gail Fleischaker for helpful correspondence and telephone conversations about various aspects of autopoiesis;
- and to Ernest Callenbach, Ed Clark, Raymond Dasmann, Leonard Duhl, Alan Miller, Stephanie Mills, and John Ryan for numerous discussions and correspondence about the principles of ecology.

During the last few years, while I was working on the book, I had several valuable opportunities to present my ideas to colleagues and students for critical discussion. I am indebted to Satish Kumar for inviting me to teach courses on "The Web of Life" at Schumacher College in England during three consecutive summers, 1992–94; and to my students in these three courses for countless critical questions and helpful suggestions. I am also grateful to Stephan Harding for teaching seminars on Gaia theory during my courses and for his generous help with numerous questions about biology and ecology. Research assistance by two of my Schumacher students, William Holloway and Morten Flatau, is also gratefully acknowledged.

In the course of my work at the Center for Ecoliteracy in Berkeley, I have had ample opportunity to discuss the characteristics of systems thinking and the principles of ecology with teachers and educators, which helped me greatly in refining my presentation of these concepts and ideas. I especially wish to thank Zenobia Barlow for organizing a series of ecoliteracy dialogues, during which most of these conversations took place.

I also had the unique opportunity of presenting various parts of the book for critical discussions in a regular series of "systems salons," held by Joanna Macy during 1993–95. I am most grateful to Joanna, and to my colleagues Tyrone Cashman and Brian Swimme, for in-depth discussions of numerous ideas at these intimate gatherings.

I wish to thank my literary agent, John Brockman, for his encouragement and for helping me formulate the initial outline of the book that he presented to my publishers.

I am very grateful to my brother, Bernt Capra, and to Trena Cleland, Stephan Harding, and William Holloway for reading the entire manuscript and giving me valuable advice and guidance. I also wish to thank John Todd and Raffi for their comments on several chapters.

My special thanks go to Julia Ponsonby for her beautiful line drawings and her patience with my repeated requests for alterations.

I am grateful to my editor Charles Conrad at Anchor Books for his enthusiasm and helpful suggestions.

Last but not least, I want to express my deep gratitude to my wife, Elizabeth, and my daughter, Juliette, for their understanding and patience during many years, when I left their company again and again to "go upstairs" for long hours of writing.

Preface

In 1944 the Austrian physicist Erwin Schrödinger wrote a short book entitled *What Is Life?* in which he advanced clear and compelling hypotheses about the molecular structure of genes. This book stimulated biologists to think about genetics in a novel way and in so doing opened a new frontier of science, molecular biology.

During subsequent decades, this new field generated a series of triumphant discoveries, culminating in the unraveling of the genetic code. However, these spectacular advances did not bring biologists any closer to answering the question posed in the title of Schrödinger's book. Nor were they able to answer the many associated questions that have puzzled scientists and philosophers for hundreds of years: How did complex structures evolve out of a random collection of molecules? What is the relationship between mind and brain? What is consciousness?

Molecular biologists have discovered the fundamental building blocks of life, but this has not helped them to understand the vital integrative actions of living organisms. Twenty-five years ago one of the leading molecular biologists, Sidney Brenner, made the following reflective comments:

> In one way, you could say all the genetic and molecular biological work of the last sixty years could be considered a long interlude. . . . Now that that program has been completed, we have come full circle—back to the problems left behind unsolved. How does a wounded organism regenerate to exactly the same structure it had before? How does the egg form the organism? . . . I think in the next twenty-five years we are going to have to teach biologists another language. . . . I don't know what it's called yet; nobody knows. . . . It may be wrong to believe that all the logic is at the molecular level. We may need to get beyond the clock mechanisms.[1]

Since the time Brenner made these comments, a new language for understanding the complex, highly integrative systems of life has indeed emerged. Different scientists call it by different names—"dynamical systems theory," "the theory of complexity," "nonlinear dynamics," "network dynamics," and so on. Chaotic attractors, fractals, dissipative structures, self-organization, and autopoietic networks are some of its key concepts.

This approach to understanding life is pursued by outstanding researchers and their teams around the world—Ilya Prigogine at the University of Brussels, Humberto Maturana at the University of Chile in Santiago, Francisco Varela at the Ecole Polytechnique in Paris, Lynn Margulis at the University of Massachusetts, Benoît Mandelbrot at Yale University, and Stuart Kauffman at the Santa Fe Institute, to name just a few. Several key discoveries of these scientists, published in technical papers and books, have been hailed as revolutionary.

However, to date nobody has proposed an overall synthesis that integrates the new discoveries into a single context and thus allows lay readers to understand them in a coherent way. This is the challenge and the promise of *The Web of Life.*

The new understanding of life may be seen as the scientific forefront of the change of paradigms from a mechanistic to an ecological worldview, which I discussed in my previous book *The Turning Point.* The present book, in a sense, is a continuation and expansion of the chapter in *The Turning Point* titled "The Systems View of Life."

The intellectual tradition of systems thinking, and the models and theories of living systems developed during the early decades of the century, form the conceptual and historical roots of the scientific framework discussed in this book. In fact, the synthesis of current theories and models I propose here may be seen as an outline of an emerging theory of living systems that offers a unified view of mind, matter, and life.

This book is for the general reader. I have kept the language as nontechnical as possible and have defined all technical terms where they first appear. However, the ideas, models, and theories I discuss are complex, and at times I felt the need to go into some technical detail to convey their substance. This applies particularly to some passages in chapters 5 and 6 and to the first part of chapter 9. Readers not interested in the technical details may want merely to browse through those passages and should feel free to skip them altogether without being afraid of losing the main thread of my argument.

The reader will also notice that the text includes not only numerous references to the literature, but also an abundance of cross-references to pages in this book. In my struggle to communicate a complex network of concepts and ideas within the linear constraints of written language, I felt that it would help to interconnect the text by a network of footnotes. My hope is that the reader will find that, like the web of life, the book itself is a whole that is more than the sum of its parts.

Berkeley, August 1995 FRITJOF CAPRA

PART ONE

The Cultural Context

1

Deep Ecology—A New Paradigm

This book is about a new scientific understanding of life at all levels of living systems—organisms, social systems, and ecosystems. It is based on a new perception of reality that has profound implications not only for science and philosophy, but also for business, politics, health care, education, and everyday life. It is therefore appropriate to begin with an outline of the broad social and cultural context of the new conception of life.

Crisis of Perception

As the century draws to a close, environmental concerns have become of paramount importance. We are faced with a whole series of global problems that are harming the biosphere and human life in alarming ways that may soon become irreversible. We have ample documentation about the extent and significance of these problems.[1]

The more we study the major problems of our time, the more we come to realize that they cannot be understood in isolation. They are systemic problems, which means that they are interconnected and interdependent. For example, stabilizing world population will be possible only when poverty is reduced worldwide.

The extinction of animal and plant species on a massive scale will continue as long as the Southern Hemisphere is burdened by massive debts. Scarcities of resources and environmental degradation combine with rapidly expanding populations to lead to the breakdown of local communities and to the ethnic and tribal violence that has become the main characteristic of the post–cold war era.

Ultimately these problems must be seen as just different facets of one single crisis, which is largely a crisis of perception. It derives from the fact that most of us, and especially our large social institutions, subscribe to the concepts of an outdated worldview, a perception of reality inadequate for dealing with our overpopulated, globally interconnected world.

There *are* solutions to the major problems of our time, some of them even simple. But they require a radical shift in our perceptions, our thinking, our values. And, indeed, we are now at the beginning of such a fundamental change of worldview in science and society, a change of paradigms as radical as the Copernican revolution. But this realization has not yet dawned on most of our political leaders. The recognition that a profound change of perception and thinking is needed if we are to survive has not yet reached most of our corporate leaders, either, or the administrators and professors of our large universities.

Not only do our leaders fail to see how different problems are interrelated; they also refuse to recognize how their so-called solutions affect future generations. From the systemic point of view, the only viable solutions are those that are "sustainable." The concept of sustainability has become a key concept in the ecology movement and is indeed crucial. Lester Brown of the Worldwatch Institute has given a simple, clear, and beautiful definition: "A sustainable society is one that satisfies its needs without diminishing the prospects of future generations."[2] This, in a nutshell, is the great challenge of our time: to create sustainable communities—that is to say, social and cultural environments in which we can satisfy our needs and aspirations without diminishing the chances of future generations.

The Paradigm Shift

My main interest in my life as a physicist has been in the dramatic change of concepts and ideas that occurred in physics during the first three decades of the century and is still being elaborated in our current theories of matter. The new concepts in physics have brought about a profound change in our worldview; from the mechanistic worldview of Descartes and Newton to a holistic, ecological view.

The new view of reality was by no means easy to accept for physicists at the beginning of the century. The exploration of the atomic and subatomic world brought them in contact with a strange and unexpected reality. In their struggle to grasp this new reality, scientists became painfully aware that their basic concepts, their language, and their whole way of thinking were inadequate to describe atomic phenomena. Their problems were not merely intellectual but amounted to an intense emotional and, one could say, even existential crisis. It took them a long time to overcome this crisis, but in the end they were rewarded with deep insights into the nature of matter and its relation to the human mind.[3]

The dramatic changes of thinking that happened in physics at the beginning of this century have been widely discussed by physicists and philosophers for more than fifty years. They led Thomas Kuhn to the notion of a scientific "paradigm," defined as "a constellation of achievements—concepts, values, techniques, etc.—shared by a scientific community and used by that community to define legitimate problems and solutions."[4] Changes of paradigms, according to Kuhn, occur in discontinuous, revolutionary breaks called "paradigm shifts."

Today, twenty-five years after Kuhn's analysis, we recognize the paradigm shift in physics as an integral part of a much larger cultural transformation. The intellectual crisis of the quantum physicists in the 1920s is mirrored today by a similar but much broader cultural crisis. Accordingly, what we are seeing is a shift of paradigms not only within science, but also in the larger social arena.[5] To analyze that cultural transformation I have generalized

Kuhn's definition of a scientific paradigm to that of a social paradigm, which I define as "a constellation of concepts, values, perceptions, and practices shared by a community, which forms a particular vision of reality that is the basis of the way the community organizes itself."[6]

The paradigm that is now receding has dominated our culture for several hundred years, during which it has shaped our modern Western society and has significantly influenced the rest of the world. This paradigm consists of a number of entrenched ideas and values, among them the view of the universe as a mechanical system composed of elementary building blocks, the view of the human body as a machine, the view of life in society as a competitive struggle for existence, the belief in unlimited material progress to be achieved through economic and technological growth, and—last, but not least—the belief that a society in which the female is everywhere subsumed under the male is one that follows a basic law of nature. All of these assumptions have been fatefully challenged by recent events. And, indeed, a radical revision of them is now occurring.

Deep Ecology

The new paradigm may be called a holistic worldview, seeing the world as an integrated whole rather than a dissociated collection of parts. It may also be called an ecological view, if the term "ecological" is used in a much broader and deeper sense than usual. Deep ecological awareness recognizes the fundamental interdependence of all phenomena and the fact that, as individuals and societies, we are all embedded in (and ultimately dependent on) the cyclical processes of nature.

The two terms "holistic" and "ecological" differ slightly in their meanings, and it seems that "holistic" is somewhat less appropriate to describe the new paradigm. A holistic view of, say, a bicycle means to see the bicycle as a functional whole and to understand the interdependence of its parts accordingly. An ecological view of the bicycle includes that, but it adds to it the perception of how the bicycle is embedded in its natural and social environment—where the raw materials that went into it came from, how it was manufactured, how its use affects the natural environment and the community by which it is used, and so on. This distinction between "holistic" and "ecological" is even more important when we talk about living systems, for which the connections with the environment are much more vital.

The sense in which I use the term "ecological" is associated with a specific philosophical school and, moreover, with a global grass-roots movement known as "deep ecology," which is rapidly gaining prominence.[7] The philosophical school was founded by the Norwegian philosopher Arne Naess in the early 1970s with his distinction between "shallow" and "deep" ecology. This distinction is now widely accepted as a very useful term for referring to a major division within contemporary environmental thought.

Shallow ecology is anthropocentric, or human-centered. It views humans as above or outside of nature, as the source of all value, and ascribes only instrumental, or "use," value to nature. Deep ecology does not separate humans—or anything else—from the natural environment. It sees the world not as a collection of isolated objects, but as a network of phenomena that are fundamentally interconnected and interdependent. Deep ecology recognizes the intrinsic value of all living beings and views humans as just one particular strand in the web of life.

Ultimately, deep ecological awareness is spiritual or religious awareness. When the concept of the human spirit is understood as the mode of consciousness in which the individual feels a sense of belonging, of connectedness, to the cosmos as a whole, it becomes clear that ecological awareness is spiritual in its deepest essence. It is, therefore, not surprising that the emerging new vision of reality based on deep ecological awareness is consistent with the so-called perennial philosophy of spiritual traditions, whether we talk about the spirituality of Christian mystics, that of Buddhists, or the philosophy and cosmology underlying the Native American traditions.[8]

There is another way in which Arne Naess has characterized deep ecology. "The essence of deep ecology," he says, "is to ask deeper questions."[9] This is also the essence of a paradigm shift.

We need to be prepared to question every single aspect of the old paradigm. Eventually we will not need to throw everything away, but before we know that we need to be willing to question everything. So deep ecology asks profound questions about the very foundations of our modern, scientific, industrial, growth-oriented, materialistic worldview and way of life. It questions this entire paradigm from an ecological perspective: from the perspective of our relationships to one another, to future generations, and to the web of life of which we are part.

Social Ecology and Ecofeminism

In addition to deep ecology, there are two other important philosophical schools of ecology, social ecology and feminist ecology, or "ecofeminism." In recent years there has been a lively debate in philosophical journals about the relative merits of deep ecology, social ecology, and ecofeminism.[10] It seems to me that each of the three schools addresses important aspects of the ecological paradigm and, rather than competing with each other, their proponents should try to integrate their approaches into a coherent ecological vision.

Deep ecological awareness seems to provide the ideal philosophical and spiritual basis for an ecological lifestyle and for environmental activism. However, it does not tell us much about the cultural characteristics and patterns of social organization that have brought about the current ecological crisis. This is the focus of social ecology.[11]

The common ground of the various schools of social ecology is the recognition that the fundamentally antiecological nature of many of our social and economic structures and their technologies is rooted in what Riane Eisler has called the "dominator system" of social organization.[12] Patriarchy, imperialism, capitalism, and racism are examples of social domination that are exploitative and antiecological. Among the different schools of social ecology there are various Marxist and anarchist groups who use their respective conceptual frameworks to analyze different patterns of social domination.

Ecofeminism could be viewed as a special school of social ecology, since it, too, addresses the basic dynamics of social domination within the context of patriarchy. However, its cultural analysis of the many facets of patriarchy and of the links between feminism and ecology go far beyond the framework of social ecology. Ecofeminists see the patriarchal domination of women by men as the prototype of all domination and exploitation in the various hierarchical, militaristic, capitalist, and industrialist forms. They point out that the exploitation of nature, in particular, has gone hand in hand with that of women, who have been identified with nature throughout the ages. This ancient association of woman and nature links women's history and the history of the environment and is the source of a natural kinship between feminism and ecology.[13] Accordingly, ecofeminists see female experiential knowledge as a major source for an ecological vision of reality.[14]

New Values

In this brief outline of the emerging ecological paradigm, I have so far emphasized the shifts in perceptions and ways of thinking. If that were all that were necessary, the transition to the new paradigm would be much easier. There are enough articulate and eloquent thinkers in the deep ecology movement who could convince our political and corporate leaders of the merits of the new thinking. But that is only part of the story. The shift of paradigms requires an expansion not only of our perceptions and ways of thinking, but also of our values.

Here it is interesting to note the striking connection in the changes between thinking and values. Both may be seen as shifts from self-assertion to integration. These two tendencies—the self-assertive and the integrative—are both essential aspects of all living systems.[15] Neither is intrinsically good or bad. What is good, or healthy, is a dynamic balance; what is bad, or unhealthy, is imbalance—overemphasis of one tendency and neglect of the other. If we now look at our Western industrial culture, we see that we have overemphasized the self-assertive and neglected the integrative tendencies. This is apparent both in our thinking and in our values, and it is very instructive to put these opposite tendencies side by side.

Thinking		***Values***	
Self-Assertive	*Integrative*	*Self-Assertive*	*Integrative*
rational	intuitive	expansion	conservation
analysis	synthesis	competition	cooperation
reductionist	holistic	quantity	quality
linear	nonlinear	domination	partnership

One of the things we notice when we look at this table is that the self-assertive values—competition, expansion, domination—are generally associated with men. Indeed, in patriarchal society they are not only favored but also given economic rewards and political power. This is one of the reasons why the shift to a more balanced value system is so difficult for most people and especially for men.

Power, in the sense of domination over others, is excessive self-assertion. The social structure in which it is exerted most effectively is the hierarchy. Indeed, our political, military, and corporate structures are hierarchically ordered, with men generally occupying the upper levels and women the lower levels. Most of these men, and quite a few women, have come to see their position in the hierarchy as part of their identity, and thus the shift to a different system of values generates existential fear in them.

However, there is another kind of power, one that is more appropriate for the new paradigm—power as influence of others. The ideal structure for exerting this kind of power is not the hierarchy but the network, which, as we shall see, is also the central metaphor of ecology.[16] The paradigm shift thus includes a shift in social organization from hierarchies to networks.

Ethics

The whole question of values is crucial to deep ecology; it is, in fact, its central defining characteristic. Whereas the old paradigm is based on anthropocentric (human-centered) values, deep ecology is grounded in ecocentric (earth-centered) values. It is a worldview that acknowledges the inherent value of nonhuman life. All living beings are members of ecological communities bound together in a network of interdependencies. When this deep ecological perception becomes part of our daily awareness, a radically new system of ethics emerges.

Such a deep ecological ethics is urgently needed today, and especially in science, since most of what scientists do is not life-furthering and life-preserving but life-destroying. With physicists designing weapons systems that threaten to wipe out life on the planet, with chemists contaminating the global environment, with biologists releasing new and unknown types of microorganisms without knowing the consequences, with psychologists and other scientists torturing animals in the name of scientific progress—with all these activities going on, it seems most urgent to introduce "ecoethical" standards into science.

It is generally not recognized that values are not peripheral to science and technology but constitute their very basis and driving force. During the scientific revolution in the seventeenth century, values were separated from facts, and ever since that time we have tended to believe that scientific facts are independent of what we do and are therefore independent of our values. In reality, scientific facts emerge out of an entire constellation of human perceptions, values, and actions—in one word, out of a paradigm—from which they cannot be separated. Although much of the detailed research may not depend explicitly on the scientist's value system, the larger paradigm within which this research is pursued will never be value free. Scientists, therefore, are responsible for their research not only intellectually but also morally.

Within the context of deep ecology, the view that values are inherent in all of living nature is grounded in the deep ecological, or spiritual, experience that nature and the self are one. This expansion of the self all the way to the identification with nature is the grounding of deep ecology, as Arne Naess clearly recognizes:

> Care flows naturally if the "self" is widened and deepened so that protection of free Nature is felt and conceived as protection of ourselves. . . . Just as we need no morals to make us breathe . . . [so] if your "self" in the wide sense embraces another being, you need no moral exhortation to show care. . . . You care for yourself without feeling any moral pressure to do it. . . . If reality is like it is experienced by the ecological self, our behavior *naturally* and beautifully follows norms of strict environmental ethics.[17]

What this implies is that the connection between an ecological perception of the world and corresponding behavior is not a logical but a *psychological* connection.[18] Logic does not lead us from the fact that we are an integral part of the web of life to certain norms of how we should live. However, if we have deep ecological awareness, or experience, of being part of the web of life, then we *will* (as opposed to *should)* be inclined to care for all of living nature. Indeed, we can scarcely refrain from responding in this way.

The link between ecology and psychology that is established by the concept of the ecological self has recently been explored by several authors. Deep ecologist Joanna Macy writes about "the greening of the self";[19] philosopher Warwick Fox has coined the term "transpersonal ecology";[20] and cultural historian Theodore Roszak uses the term "eco-psychology"[21] to express the deep connection between these two fields, which until very recently were completely separate.

Shift from Physics to the Life Sciences

By calling the emerging new vision of reality "ecological" in the sense of deep ecology, we emphasize that life is at its very center. This is an important point for science, because in the old paradigm physics has been the model and source of metaphors for all other sciences. "All philosophy is like a tree," wrote Descartes. "The roots are metaphysics, the trunk is physics, and the branches are all the other sciences."[22]

Deep ecology has overcome this Cartesian metaphor. Even though the paradigm shift in physics is still of special interest because it was the first to occur in modern science, physics has now lost its role as the science providing the most fundamental description of reality. However, this is still not generally recognized today. Scientists as well as nonscientists frequently retain the popular belief that "if you really want to know the ultimate explanation, you have to ask a physicist," which is clearly a Cartesian fallacy. Today the paradigm shift in science, at its deepest level, implies a shift from physics to the life sciences.

PART TWO

The Rise of Systems Thinking

2

From the Parts to the Whole

During this century the change from the mechanistic to the ecological paradigm has proceeded in different forms and at different speeds in the various scientific fields. It is not a steady change. It involves scientific revolutions, backlashes, and pendulum swings. A chaotic pendulum in the sense of chaos theory[1]—oscillations that almost repeat themselves, but not quite, seemingly random and yet forming a complex, highly organized pattern—would perhaps be the most appropriate contemporary metaphor.

The basic tension is one between the parts and the whole. The emphasis on the parts has been called mechanistic, reductionist, or atomistic; the emphasis on the whole holistic, organismic, or ecological. In twentieth-century science the holistic perspective has become known as "systemic" and the way of thinking it implies as "systems thinking." In this book I shall use "ecological" and "systemic" synonymously, "systemic" being merely the more technical, scientific term.

The main characteristics of systems thinking emerged simultaneously in several disciplines during the first half of the century, especially during the 1920s. Systems thinking was pioneered by biologists, who emphasized the view of living organisms as integrated wholes. It was further enriched by Gestalt psychology and the new science of ecology, and it had perhaps the most dramatic effects in quantum physics. Since the central idea of the new paradigm concerns the nature of life, let us first turn to biology.

Substance and Form

The tension between mechanism and holism has been a recurring theme throughout the history of biology. It is an inevitable consequence of the ancient dichotomy between substance (matter, structure, quantity) and form (pattern, order, quality). Biological form is more than shape, more than a static configuration of components in a whole. There is a continual flux of matter through a living organism, while its form is maintained. There is development, and there is evolution. Thus the understanding of biological form is inextricably linked to the understanding of metabolic and developmental processes.

At the dawn of Western philosophy and science, the Pythagoreans distinguished "number," or pattern, from substance, or matter, viewing it as something that limits matter and gives it shape. As Gregory Bateson put it:

> The argument took the shape of "Do you ask what it's made of—earth, fire, water, etc.?" Or do you ask, "What is its *pattern?*" Pythagoreans stood for inquiring into pattern rather than inquiring into substance.[2]

Aristotle, the first biologist in the Western tradition, also distinguished between matter and form but at the same time linked the two through a process of development.[3] In contrast with Plato, Aristotle believed that form had no separate existence but was immanent in matter. Nor could matter exist separately from form. Matter, according to Aristotle, contains the essential nature of all things, but only as potentiality. By means of form this essence becomes real, or actual. The process of the self-realization of the essence in the actual phenomena is by Aristotle called *entelechy* ("self-completion"). It is a process of development, a thrust toward full self-realization. Matter and form are the two sides of this process, separable only through abstraction.

Aristotle created a formal system of logic and a set of unifying concepts, which he applied to the main disciplines of his time—biology, physics, metaphysics, ethics, and politics. His philosophy and science dominated Western thought for two thousand years after his death, during which his authority became almost as unquestioned as that of the church.

Cartesian Mechanism

In the sixteenth and seventeenth centuries the medieval worldview, based on Aristotelian philosophy and Christian theology, changed radically. The notion of an organic, living, and spiritual universe was replaced by that of the world as a machine, and the world machine became the dominant metaphor of the modern era. This radical change was brought about by the new discoveries in physics, astronomy, and mathematics known as the Scientific Revolution and associated with the names of Copernicus, Galileo, Descartes, Bacon, and Newton.[4]

Galileo Galilei banned quality from science, restricting it to the study of phenomena that could be measured and quantified. This has been a very successful strategy throughout modern science, but our obsession with quantification and measurement has also exacted a heavy toll. As the psychiatrist R. D. Laing put it emphatically:

> Galileo's program offers us a dead world: Out go sight, sound, taste, touch, and smell, and along with them have since gone esthetic and ethical sensibility, values, quality, soul, consciousness, spirit. Experience as such is cast out of the realm of scientific discourse. Hardly anything has changed our world more during the past four hundred years than Galileo's audacious program. We had to destroy the world in theory before we could destroy it in practice.[5]

René Descartes created the method of analytic thinking, which consists in breaking up complex phenomena into pieces to understand the behavior of the whole from the properties of its parts. Descartes based his view of nature on the fundamental division between two independent and separate realms—that of mind and that of matter. The material universe, including living organisms, was a machine for Descartes, which could in principle be understood completely by analyzing it in terms of its smallest parts.

The conceptual framework created by Galileo and Descartes—the world as a perfect machine governed by exact mathematical laws—was completed triumphantly by Isaac Newton, whose grand synthesis, Newtonian mechanics, was the crowning achievement of seventeenth-century science. In biology the greatest success of Descartes's mechanistic model was its application to the phenomenon of blood circulation by William Harvey. Inspired by Harvey's success, the physiologists of his time tried to apply the mechanistic method to describe other bodily functions, such as digestion and metabolism. These attempts were dismal failures, however, because the phenomena the physiologists tried to explain involved chemical processes that were unknown at the time and could not be described in mechanical terms. The situation changed significantly in the eighteenth century, when Antoine Lavoisier, the "father of modern chemistry," demonstrated that respiration is a special form of oxidation and thus confirmed the relevance of chemical processes to the functioning of living organisms.

In the light of the new science of chemistry, the simplistic mechanical models of living organisms were largely abandoned, but the essence of the Cartesian idea survived. Animals were still machines, although they were much more complicated than mechanical clockworks, involving complex chemical processes. Accordingly, Cartesian mechanism was expressed in the dogma that the laws of biology can ultimately be reduced to those of physics and chemistry. At the same time, the rigidly mechanistic physiology found its most forceful and elaborate expression in a polemic treatise *Man a Machine,* by Julien de La Mettrie, which remained famous well beyond the eighteenth century and generated many debates and controversies, some of which reached even into the twentieth century.[6]

The Romantic Movement

The first strong opposition to the mechanistic Cartesian paradigm came from the Romantic movement in art, literature, and philosophy in the late eighteenth and nineteenth centuries. William Blake, the great mystical poet and painter who exerted a strong influence on English Romanticism, was a passionate critic of Newton. He summarized his critique in these celebrated lines:

> May God us keep
> from single vision and Newton's sleep.[7]

The German Romantic poets and philosophers returned to the Aristotelian tradition by concentrating on the nature of organic form. Goethe, the central figure in this movement, was among the first to use the term "morphology" for the study of biological form from a dynamic, developmental point of view. He admired nature's "moving order" *(bewegliche Ordnung)* and conceived of form as a pattern of relationships within an organized whole—a conception that is at the forefront of contemporary systems thinking. "Each creature," wrote Goethe, "is but a patterned gradation *(Schattierung)* of one great harmonious whole."[8] The Romantic artists were concerned mainly with a qualitative understanding of patterns, and therefore they placed great emphasis on explaining the basic properties of life in terms of visualized forms. Goethe, in particular, felt that visual perception was the door to understanding organic form.[9]

The understanding of organic form also played an important role in the philosophy of Immanuel Kant, who is often considered the greatest of the modern philosophers. An idealist, Kant separated the phenomenal world from a world of "things-in-themselves." He believed that science could offer only mechanical explanations, but he affirmed that in areas where such explanations were inadequate, scientific knowledge needed to be supplemented by considering nature as being purposeful. The most important of these areas, according to Kant, is the understanding of life.[10]

In his *Critique of Judgment* Kant discussed the nature of living organisms. He argued that organisms, in contrast with machines, are self-reproducing, self-organizing wholes. In a machine, according to Kant, the parts only exist *for* each other, in the sense of supporting each other within a functional whole. In an organism the parts also exist *by means of* each other, in the sense of producing one another.[11] "We must think of each part as an organ," wrote Kant, "that produces the other parts (so that each reciprocally produces the other). . . . Because of this, [the organism] will be both an organized and self-organizing being."[12] With this statement Kant became not only the first to use the term "self-organization" to define the nature of living organisms, he also used it in a way that is remarkably similar to some contemporary conceptions.[13]

The Romantic view of nature as "one great harmonious whole," as Goethe put it, led some scientists of that period to extend their search for wholeness to the entire planet and see the Earth as an integrated whole, a living being. The view of the Earth as being alive, of course, has a long tradition. Mythical images of the Earth Mother are among the oldest in human religious history. Gaia, the Earth Goddess, was revered as the supreme deity in early, pre-Hellenic Greece.[14] Earlier still, from the Neolithic through the Bronze Ages, the societies of "Old Europe" worshiped numerous female deities as incarnations of Mother Earth.[15]

The idea of the Earth as a living, spiritual being continued to flourish throughout the Middle Ages and the Renaissance, until the whole medieval outlook was replaced by the Cartesian image of the world as a machine. So when scientists in the eighteenth century began to visualize the Earth as a living being, they revived an ancient tradition that had been dormant for only a relatively brief period.

More recently, the idea of a living planet was formulated in modern scientific language as the so-called Gaia hypothesis, and it is interesting that the views of the living Earth developed by eighteenth-century scientists contain some key elements of our contemporary theory.[16] The Scottish geologist James Hutton maintained that geological and biological processes are all interlinked and compared the Earth's waters to the circulatory system of an animal. The German naturalist and explorer Alexander von Humboldt, one of the greatest unifying thinkers of the eighteenth and nineteenth centuries, took this idea even further. His "habit of viewing the Globe as a great whole" led Humboldt to identifying climate as a unifying global force and to recognizing the coevolution of living organisms, climate, and Earth crust, which almost encapsulates the contemporary Gaia hypothesis.[17]

At the end of the eighteenth and the beginning of the nineteenth centuries the influence of the Romantic movement was so strong that the primary concern of biologists was the problem of biological form, and questions of material composition were secondary. This was especially true for the great French schools of comparative anatomy, or "morphology," pioneered by Georges Cuvier, who created a system of zoological classification based on similarities of structural relations.[18]

Nineteenth-Century Mechanism

During the second half of the nineteenth century the pendulum swung back to mechanism, when the newly perfected microscope led to many remarkable advances in biology.[19] The nineteenth century is best known for the establishment of evolutionary thought, but it also saw the formulation of cell theory, the beginning of modern embryology, the rise of microbiology, and the discovery of the laws of heredity. These new discoveries grounded biology firmly in physics and chemistry, and scientists renewed their efforts to search for physico-chemical explanations of life.

When Rudolf Virchow formulated cell theory in its modern form, the focus of biologists shifted from organisms to cells. Biological functions, rather than reflecting the organization of the organism as a whole, were now seen as the results of interactions among the cellular building blocks.

Research in microbiology—a new field that revealed an unsuspected richness and complexity of microscopic living organisms—was dominated by the genius of Louis Pasteur, whose penetrating insights and clear formulations made a lasting impact in chemistry, biology, and medicine. Pasteur was able to establish the role of bacteria in certain chemical processes, thus laying the foundations of the new science of biochemistry, and he demonstrated that there is a definite correlation between "germs" (microorganisms) and disease.

Pasteur's discoveries led to a simplistic "germ theory of disease," in which bacteria were seen as the only cause of disease. This reductionist view eclipsed an alternative theory that had been taught a few years earlier by Claude Bernard, the founder of modern experimental medicine. Bernard insisted on the close and intimate relation between an organism and its environment and was the first to point out that each organism also has an internal environment, in which its organs and tissues live. Bernard observed that in a healthy organism this internal environment remains essentially constant, even when the external environment fluctuates considerably. His concept of the constancy of the internal environment foreshadowed the important notion of homeostasis, developed by Walter Cannon in the 1920s.

The new science of biochemistry progressed steadily and established the firm belief among biologists that all properties and functions of living organisms would eventually be explained in terms of chemical and physical laws. This belief was most clearly expressed by Jacques Loeb in *The Mechanistic Conception of Life,* which had a tremendous influence on the biological thinking of its time.

Vitalism

The triumphs of nineteenth-century biology—cell theory, embryology, and microbiology—established the mechanistic conception of life as a firm dogma among biologists. Yet they carried within themselves the seeds of the next wave of opposition, the school known as organismic biology, or "organicism." While cell biology made enormous progress in understanding the structures and functions of many of the cell's subunits, it remained largely ignorant of the coordinating activities that integrate those operations into the functioning of the cell as a whole.

The limitations of the reductionist model were shown even more dramatically by the problems of cell development and differentiation. In the very early stages of the development of higher organisms, the number of their cells increases from one to two, to four, and so forth, doubling at each step. Since the genetic information is identical in each cell, how can these cells specialize in different ways, becoming muscle cells, blood cells, bone cells, nerve cells, and so on? This basic problem of development, which appears in many variations throughout biology, clearly flies in the face of the mechanistic view of life.

Before organicism was born, many outstanding biologists went through a phase of vitalism, and for many years the debate between mechanism and holism was framed as one between mechanism and vitalism.[20] A clear understanding of the vitalist idea is very useful, since it stands in sharp contrast with the systems view of life that was to emerge from organismic biology in the twentieth century.

Vitalism and organicism are both opposed to the reduction of biology to physics and chemistry. Both schools maintain that although the laws of physics and chemistry are applicable to organisms, they are insufficient to fully understand the phenomenon of life. The behavior of a living organism as an integrated whole cannot be understood from the study of its parts alone. As the systems theorists would put it several decades later, the whole is more than the sum of its parts.

Vitalists and organismic biologists differ sharply in their answers to the question In what sense exactly is the whole more than the sum of its parts? Vitalists assert that some nonphysical entity, force, or field must be added to the laws of physics and chemistry to understand life. Organismic biologists maintain that the additional ingredient is the understanding of "organization," or "organizing relations."

Since these organizing relations are patterns of relationships immanent in the physical structure of the organism, organismic biologists assert that no separate, nonphysical entity is required for the understanding of life. We shall see later on that the concept of organization has been refined to that of "self-organization" in contemporary theories of living systems and that understanding the pattern of self-organization is the key to understanding the essential nature of life.

Whereas organismic biologists challenged the Cartesian machine analogy by trying to understand biological form in terms of a wider meaning of organization, vitalists did not really go beyond the Cartesian paradigm. Their language was limited by the same images and metaphors; they merely added a nonphysical entity as the designer or director of the organizing processes that defy mechanistic explanations. Thus the Cartesian split of mind and body led to both mechanism and vitalism. When Descartes's followers banned the mind from biology and conceived the body as a machine, the "ghost in the machine"—to use Arthur Koestler's phrase[21]—soon reappeared in vitalist theories.

The German embryologist Hans Driesch initiated the opposition to mechanistic biology at the turn of the century with his pioneering experiments on sea urchin eggs, which led him to formulate the first theory of vitalism. When Driesch destroyed one of the cells of an embryo at the very early two-celled stage, the remaining cell developed not into half a sea urchin, but into a complete but smaller organism. Similarly, complete smaller organisms developed after the destruction of two or three cells in four-celled embryos. Driesch realized that his sea urchin eggs had done what a machine could never do: they had regenerated wholes from some of their parts.

To explain this phenomenon of self-regulation, Driesch seems to have looked strenuously for the missing pattern of organization.[22] But instead of turning to the concept of pattern, he postulated a causal factor, for which he chose the Aristotelian term *entelechy*. However, whereas Aristotle's *entelechy* is a process of self-realization that unifies matter and form, the *entelechy* postulated by Driesch is a separate entity, acting on the physical system without being part of it.

The vitalist idea has been revived recently in much more sophisticated form by Rupert Sheldrake, who postulates the existence of nonphysical *morphogenetic* ("form-generating") fields as the causal agents of the development and maintenance of biological form.[23]

Organismic Biology

During the early twentieth century organismic biologists, opposing both mechanism and vitalism, took up the problem of biological form with new enthusiasm, elaborating and refining many of the key insights of Aristotle, Goethe, Kant, and Cuvier. Some of the main characteristics of what we now call systems thinking emerged from their extensive reflections.[24]

Ross Harrison, one of the early exponents of the organismic school, explored the concept of organization, which had gradually come to replace the old notion of function in physiology. This shift from function to organization represents a shift from mechanistic to systemic thinking, because function is essentially a mechanistic concept. Harrison identified configuration and relationship as two important aspects of organization, which were subsequently unified in the concept of pattern as a configuration of ordered relationships.

The biochemist Lawrence Henderson was influential through his early use of the term "system" to denote both living organisms and social systems.[25] From that time on, a system has come to mean an integrated whole whose essential properties arise from the relationships between its parts, and "systems thinking" the understanding of a phenomenon within the context of a larger whole. This is, in fact, the root meaning of the word "system," which derives from the Greek *synhistanai* ("to place together"). To understand things systemically literally means to put them into a context, to establish the nature of their relationships.[26]

The biologist Joseph Woodger asserted that organisms could be described completely in terms of their chemical elements, "plus organizing relations." This formulation had considerable influence on Joseph Needham, who maintained that the publication of Woodger's *Biological Principles* in 1936 marked the end of the debate between mechanists and vitalists.[27] Needham, whose early work was on problems in the biochemistry of development, was always deeply interested in the philosophical and historical dimensions of science. He wrote many essays in defense of the mechanistic paradigm but subsequently came to embrace the organismic outlook. "A logical analysis of the concept of organism," he wrote in 1935, "leads us to look for organizing relations at all levels, higher and lower, coarse and fine, of the living structure."[28] Later on Needham left biology to become one of the leading historians of Chinese science and, as such, an ardent advocate of the organismic worldview that is the basis of Chinese thought.

Woodger and many others emphasized that one of the key characteristics of the organization of living organisms was its hierarchical nature. Indeed, an outstanding property of all life is the tendency to form multileveled structures of systems within systems. Each of these forms a whole with respect to its parts while at the same time being a part of a larger whole. Thus cells combine to form tissues, tissues to form organs, and organs to form organisms. These in turn exist within social systems and ecosystems. Throughout the living world we find living systems nesting within other living systems.

Since the early days of organismic biology these multileveled structures have been called hierarchies. However, this term can be rather misleading, since it is derived from human hierarchies, which are fairly rigid structures of domination and control, quite unlike the multileveled order found in nature. We shall see that the important concept of the network—the web of life—provides a new perspective on the so-called hierarchies of nature.

What the early systems thinkers recognized very clearly is the existence of different levels of complexity with different kinds of laws operating at each level. Indeed, the concept of "organized complexity" became the very subject of the systems approach.[29] At each level of complexity the observed phenomena exhibit properties that do not exist at the lower level. For example, the concept of temperature, which is central to thermodynamics, is meaningless at the level of individual atoms, where the laws of quantum theory operate. Similarly, the taste of sugar is not present in the carbon, hydrogen, and oxygen atoms that constitute its components. In the early 1920s the philosopher C. D. Broad coined the term "emergent properties" for those properties that emerge at a certain level of complexity but do not exist at lower levels.

Systems Thinking

The ideas set forth by organismic biologists during the first half of the century helped to give birth to a new way of thinking—"systems thinking"—in terms of connectedness, relationships, context. According to the systems view, the essential properties of an organism, or living system, are properties of the whole, which none of the parts have. They arise from the interactions and relationships among the parts. These properties are destroyed when the system is dissected, either physically or theoretically, into isolated elements. Although we can discern individual parts in any system, these parts are not isolated, and the nature of the whole is always different from the mere sum of its parts. The systems view of life is illustrated beautifully and abundantly in the writings of Paul Weiss, who brought systems concepts to the life sciences from his earlier studies of engineering and spent his whole life exploring and advocating a full organismic conception of biology.[30]

The emergence of systems thinking was a profound revolution in the history of Western scientific thought. The belief that in every complex system the behavior of the whole can be understood entirely from the properties of its parts is central to the Cartesian paradigm. This was Descartes's celebrated method of analytic thinking, which has been an essential characteristic of modern scientific thought. In the analytic, or reductionist, approach, the parts themselves cannot be analyzed any further, except by reducing them to still smaller parts. Indeed, Western science has been progressing in that way, and at each step there has been a level of fundamental constituents that could not be analyzed any further.

The great shock of twentieth-century science has been that systems cannot be understood by analysis. The properties of the parts are not intrinsic properties but can be understood only within the context of the larger whole. Thus the relationship between the parts and the whole has been reversed. In the systems approach the properties of the parts can be understood only from the organization of the whole. Accordingly, systems thinking concentrates not on basic building blocks, but on basic principles of organization. Systems thinking is "contextual," which is the opposite of analytical thinking. Analysis means taking something apart in order to understand it; systems thinking means putting it into the context of a larger whole.

Quantum Physics

The realization that systems are integrated wholes that cannot be understood by analysis was even more shocking in physics than in biology. Ever since Newton, physicists had believed that all physical phenomena could be reduced to the properties of hard and solid material particles. In the 1920s, however, quantum theory forced them to accept the fact that the solid material objects of classical physics dissolve at the subatomic level into wavelike patterns of probabilities. These patterns, moreover, do not represent probabilities of things, but rather probabilities of interconnections. The subatomic particles have no meaning as isolated entities but can be understood only as interconnections, or correlations, among various processes of observation and measurement. In other words, subatomic particles are not "things" but interconnections among things, and these, in turn, are interconnections among other things, and so on. In quantum theory we never end up with any "things"; we always deal with interconnections.

This is how quantum physics shows that we cannot decompose the world into independently existing elementary units. As we shift our attention from macroscopic objects to atoms and subatomic particles, nature does not show us any isolated building blocks, but rather appears as a complex web of relationships among the various parts of a unified whole. As Werner Heisenberg, one of the founders of quantum theory, put it, "The world thus appears as a complicated tissue of events, in which connections of different kinds alternate or overlap or combine and thereby determine the texture of the whole."[31]

Molecules and atoms—the structures described by quantum physics—consist of components. However, these components, the subatomic particles, cannot be understood as isolated entities b.. must be defined through their interrelations. In the words of Henry Stapp, "An elementary particle is not an independently existing unanalyzable entity. It is, in essence, a set of relationships that reach outward to other things."[32]

In the formalism of quantum theory these relationships are expressed in terms of probabilities, and the probabilities are determined by the dynamics of the whole system. Whereas in classical mechanics the properties and behavior of the parts determine those of the whole, the situation is reversed in quantum mechanics: it is the whole that determines the behavior of the parts.

During the 1920s the quantum physicists struggled with the same conceptual shift from the parts to the whole that gave rise to the school of organismic biology. In fact, the biologists would probably have found it much harder to overcome Cartesian mechanism had it not broken down in such a spectacular fashion in physics, which had been the great triumph of the Cartesian paradigm for three centuries. Heisenberg saw the shift from the parts to the whole as the central aspect of that conceptual revolution, and he was so impressed by it that he titled his scientific autobiography *Der Teil und das Ganze (The Part and the Whole).*[33]

Gestalt Psychology

When the first organismic biologists grappled with the problem of organic form and debated the relative merits of mechanism and vitalism, German psychologists contributed to that dialogue from the very beginning.[34] The German word for organic form is *Gestalt* (as distinct from *Form,* which denotes inanimate form), and the much discussed problem of organic form was known as the *Gestaltproblem* in those days. At the turn of the century, the philosopher Christian von Ehrenfels was the first to use *Gestalt* in the sense of an irreducible perceptual pattern, which sparked the school of Gestalt psychology. Ehrenfels characterized a gestalt by asserting that the whole is more than the sum of its parts, which would become the key formula of systems thinkers later on.[35]

Gestalt psychologists, led by Max Wertheimer and Wolfgang

he existence of irreducible wholes as a key aspect of iving organisms, they asserted, perceive things not in ed elements, but as integrated perceptual patterns— ganized wholes, which exhibit qualities that are ab- parts. The notion of pattern was always implicit in the writings of the Gestalt psychologists, who often used the analogy of a musical theme that can be played in different keys without losing its essential features.

Like the organismic biologists, Gestalt psychologists saw their school of thought as a third way beyond mechanism and vitalism. The Gestalt school made substantial contributions to psychology, especially in the study of learning and the nature of associations. Several decades later, during the 1960s, the holistic approach to psychology gave rise to a corresponding school of psychotherapy known as Gestalt therapy, which emphasizes the integration of personal experiences into meaningful wholes.[36]

In the Germany of the 1920s, the Weimar Republic, both organismic biology and Gestalt psychology were part of a larger intellectual trend that saw itself as a protest movement against the increasing fragmentation and alienation of human nature. The entire Weimar culture was characterized by an antimechanistic outlook, a "hunger for wholeness."[37] Organismic biology, Gestalt psychology, ecology, and, later on, general systems theory all grew out of this holistic zeitgeist.

Ecology

While organismic biologists encountered irreducible wholeness in organisms, quantum physicists in atomic phenomena, and Gestalt psychologists in perception, ecologists encountered it in their studies of animal and plant communities. The new science of ecology emerged out of the organismic school of biology during the nineteenth century, when biologists began to study communities of organisms.

Ecology—from the Greek *oikos* ("household")—is the study of the Earth Household. More precisely it is the study of the relationships that interlink all members of the Earth Household. The

term was coined in 1866 by the German biologist Erns
who defined it as "the science of relations between th
and the surrounding outer world."[38] In 1909 the w
("environment") was used for the first time by the Baltic
and ecological pioneer Jakob von Uexküll.[39] In the 1920s ecologists focused on functional relationships within animal and plant communities.[40] In his pioneering book, *Animal Ecology,* Charles Elton introduced the concepts of food chains and food cycles, viewing the feeding relationships within biological communities as their central organizing principle.

Since the language of the early ecologists was very close to that of organismic biology, it is not surprising that they compared biological communities to organisms. For example, Frederic Clements, an American plant ecologist and pioneer in the study of succession, viewed plant communities as "superorganisms." This concept sparked a lively debate, which went on for more than a decade until the British plant ecologist A. G. Tansley rejected the notion of superorganisms and coined the term "ecosystem" to characterize animal and plant communities. The ecosystem concept—defined today as "a community of organisms and their physical environment interacting as an ecological unit"[41]—shaped all subsequent ecological thinking and, by its very name, fostered a systems approach to ecology.

The term "biosphere" was first used in the late nineteenth century by the Austrian geologist Eduard Suess to describe the layer of life surrounding the Earth. A few decades later the Russian geochemist Vladimir Vernadsky developed the concept into a full-fledged theory in his pioneering book, *Biosphere.*[42] Building on the ideas of Goethe, Humboldt, and Suess, Vernadsky saw life as a "geological force" that partly creates and partly controls the planetary environment. Among all the early theories of the living Earth, Vernadsky's comes closest to the contemporary Gaia theory developed by James Lovelock and Lynn Margulis in the 1970s.[43]

The new science of ecology enriched the emerging systemic way of thinking by introducing two new concepts—community and network. By viewing an ecological community as an assemblage of organisms, bound into a functional whole by their mutual relationships, ecologists facilitated the change of focus from organisms to communities and back, applying the same kinds of concepts to different systems levels.

Today we know that most organisms are not only members of ecological communities but are also complex ecosystems themselves, containing a host of smaller organisms that have considerable autonomy and yet are integrated harmoniously into the functioning of the whole. So there are three kinds of living systems—organisms, parts of organisms, and communities of organisms—all of which are integrated wholes whose essential properties arise from the interactions and interdependence of their parts.

Over billions of years of evolution many species have formed such tightly knit communities that the whole system resembles a large, multicreatured organism.[44] Bees and ants, for example, are unable to survive in isolation, but in great numbers they act almost like the cells of a complex organism with a collective intelligence and capabilities for adaptation far superior to those of its individual members. Similar close coordination of activities exists also among different species, where it is known as symbiosis, and again the resulting living systems have the characteristics of single organisms.[45]

From the beginning of ecology, ecological communities have been seen as consisting of organisms linked together in network fashion through feeding relations. This idea is found repeatedly in the writings of nineteenth-century naturalists, and when food chains and food cycles began to be studied in the 1920s, these concepts were soon expanded to the contemporary concept of food webs.

The "web of life" is, of course, an ancient idea, which has been used by poets, philosophers, and mystics throughout the ages to convey their sense of the interwovenness and interdependence of all phenomena. One of the most beautiful expressions is found in the celebrated speech attributed to Chief Seattle, which serves as the motto for this book.

As the network concept became more and more prominent in ecology, systemic thinkers began to use network models at all systems levels, viewing organisms as networks of cells, organs, and organ systems, just as ecosystems are understood as networks of individual organisms. Correspondingly, the flows of matter and energy through ecosystems were perceived as the continuation of the metabolic pathways through organisms.

The view of living systems as networks provides a novel perspective on the so-called hierarchies of nature.[46] Since living systems at all levels are networks, we must visualize the web of life as living systems (networks) interacting in network fashion with other systems (networks). For example, we can picture an ecosystem schematically as a network with a few nodes. Each node represents an organism, which means that each node, when magnified, appears itself as a network. Each node in the new network may represent an organ, which in turn will appear as a network when magnified, and so on.

In other words, the web of life consists of networks within networks. At each scale, under closer scrutiny, the nodes of the network reveal themselves as smaller networks. We tend to arrange these systems, all nesting within larger systems, in a hierarchical scheme by placing the larger systems above the smaller ones in pyramid fashion. But this is a human projection. In nature there is no "above" or "below," and there are no hierarchies. There are only networks nesting within other networks.

During the last few decades the network perspective has become more and more central to ecology. As the ecologist Bernard Patten put it in his concluding remarks to a recent conference on ecological networks: "Ecology *is* networks. . . . To understand ecosystems ultimately will be to understand networks."[47] Indeed, during the second half of the century the network concept has been the key to the recent advances in the scientific understanding not only of ecosystems but of the very nature of life.

3

Systems Theories

By the 1930s most of the key criteria of systems thinking had been formulated by organismic biologists, Gestalt psychologists, and ecologists. In all these fields the exploration of living systems—organisms, parts of organisms, and communities of organisms—had led scientists to the same new way of thinking in terms of connectedness, relationships, and context. This new thinking was also supported by the revolutionary discoveries in quantum physics in the realm of atoms and subatomic particles.

Criteria of Systems Thinking

It is perhaps worthwhile to summarize the key characteristics of systems thinking at this point. The first, and most general, criterion is the shift from the parts to the whole. Living systems are integrated wholes whose properties cannot be reduced to those of smaller parts. Their essential, or "systemic," properties are properties of the whole, which none of the parts have. They arise from the "organizing relations" of the parts—that is, from a configuration of ordered relationships that is characteristic of that particular class of organisms, or systems. Systemic properties are destroyed when a system is dissected into isolated elements.

Another key criterion of systems thinking is the ability to shift one's attention back and forth between systems levels. Throughout the living world we find systems nesting within other systems, and by applying the same concepts to different systems levels—for example, the concept of stress to an organism, a city, or an economy—we can often gain important insights. On the other hand, we also have to recognize that, in general, different systems levels represent levels of differing complexity. At each level the observed phenomena exhibit properties that do not exist at lower levels. The systemic properties of a particular level are called "emergent" properties, since they emerge at that particular level.

In the shift from mechanistic thinking to systems thinking, the relationship between the parts and the whole has been reversed. Cartesian science believed that in any complex system the behavior of the whole could be analyzed in terms of the properties of its parts. Systems science shows that living systems cannot be understood by analysis. The properties of the parts are not intrinsic properties but can be understood only within the context of the larger whole. Thus systems thinking is "contextual" thinking; and since explaining things in terms of their context means explaining them in terms of their environment, we can also say that all systems thinking is environmental thinking.

Ultimately—as quantum physics showed so dramatically—there are no parts at all. What we call a part is merely a pattern in an inseparable web of relationships. Therefore the shift from the parts to the whole can also be seen as a shift from objects to relationships. In a sense, this is a figure/ground shift. In the mechanistic view the world is a collection of objects. These, of course, interact with one another, and hence there are relationships among them. But the relationships are secondary, as illustrated schematically below in figure 3-1A. In the systems view we realize that the objects themselves are networks of relationships, embedded in larger networks. For the systems thinker the relationships are primary. The boundaries of the discernible patterns ("objects") are secondary, as pictured—again in greatly simplified fashion—in figure 3-1B.

The perception of the living world as a network of relationships

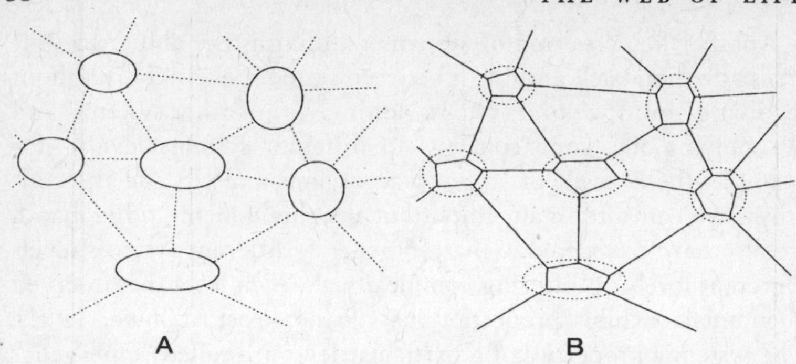

Figure 3-1
Figure/ground shift from objects to relationships.

has made thinking in terms of networks—expressed more elegantly in German as *vernetztes Denken*—another key characteristic of systems thinking. This "network thinking" has influenced not only our view of nature but also the way we speak about scientific knowledge. For thousands of years Western scientists and philosophers have used the metaphor of knowledge as a building, together with many other architectural metaphors derived from it.[1] We speak of *fundamental* laws, *fundamental* principles, *basic building blocks,* and the like, and we assert that the *edifice* of science must be built on firm *foundations.* Whenever major scientific revolutions occurred, it was felt that the foundations of science were moving. Thus Descartes wrote in his celebrated *Discourse on Method:*

> In so far as [the sciences] borrow their principles from philosophy, I considered that nothing solid could be built on such shifting foundations.[2]

Three hundred years later Heisenberg wrote in his *Physics and Philosophy* that the foundations of classical physics, that is, of the very edifice Descartes had built, were shifting:

> The violent reaction to the recent development of modern physics can only be understood when one realizes that here the foundations of physics have started moving; and that this motion has

> caused the feeling that the ground would be cut from under science.[3]

Einstein, in his autobiography, described his feelings in terms very similar to Heisenberg's:

> It was as if the ground had been pulled out from under one, with no firm foundation to be seen anywhere, upon which one could have built.[4]

In the new systems thinking, the metaphor of knowledge as a building is being replaced by that of the network. As we perceive reality as a network of relationships, our descriptions, too, form an interconnected network of concepts and models in which there are no foundations. For most scientists such a view of knowledge as a network with no firm foundations is extremely unsettling, and today it is by no means generally accepted. But as the network approach expands throughout the scientific community, the idea of knowledge as a network will undoubtedly find increasing acceptance.

The notion of scientific knowledge as a network of concepts and models, in which no part is any more fundamental than the others, was formalized in physics by Geoffrey Chew in his "bootstrap philosophy" in the 1970s.[5] The bootstrap philosophy not only abandons the idea of fundamental building blocks of matter, it accepts no fundamental entities whatsoever—no fundamental constants, laws, or equations. The material universe is seen as a dynamic web of interrelated events. None of the properties of any part of this web is fundamental; they all follow from the properties of the other parts, and the overall consistency of their interrelations determines the structure of the entire web.

When this approach is applied to science as a whole, it implies that physics can no longer be seen as the most fundamental level of science. Since there are no foundations in the network, the phenomena described by physics are not any more fundamental than those described by, say, biology or psychology. They belong to different systems levels, but none of those levels is any more fundamental than the others.

Another important implication of the view of reality as an inseparable network of relationships concerns the traditional concept of scientific objectivity. In the Cartesian paradigm scientific descriptions are believed to be objective—that is, independent of the human observer and the process of knowing. The new paradigm implies that epistemology—understanding of the process of knowing—has to be included explicitly in the description of natural phenomena.

This recognition entered into science with Werner Heisenberg and is closely related to the view of physical reality as a web of relationships. If we imagine the network pictured previously in figure 3-1B as much more intricate, perhaps somewhat similar to an inkblot in a Rorschach test, we can easily understand that isolating a pattern in this complex network by drawing a boundary around it and calling it an "object" will be somewhat arbitrary.

Indeed, this is what happens when we refer to objects in our environment. For example, when we see a network of relationships among leaves, twigs, branches, and a trunk, we call it a "tree." When we draw a picture of a tree, most of us will not draw the roots. Yet the roots of a tree are often as expansive as the parts we see. In a forest, moreover, the roots of all trees are interconnected and form a dense underground network in which there are no precise boundaries between individual trees.

In short, what we call a tree depends on our perceptions. It depends, as we say in science, on our methods of observation and measurement. In the words of Heisenberg: "What we observe is not nature itself, but nature exposed to our method of questioning."[6] Thus systems thinking involves a shift from objective to "epistemic" science, to a framework in which epistemology—"the method of questioning"—becomes an integral part of scientific theories.

The criteria of systems thinking described in this brief summary are all interdependent. Nature is seen as an interconnected web of relationships, in which the identification of specific patterns as "objects" depends on the human observer and the process of knowing. This web of relationships is described in terms of a corresponding network of concepts and models, none of which is any more fundamental than the others.

This new approach to science immediately raises an important question. If everything is connected to everything else, how can we ever hope to understand anything? Since all natural phenomena are ultimately interconnected, in order to explain any one of them we need to understand all the others, which is obviously impossible.

What makes it possible to turn the systems approach into a science is the discovery that there is approximate knowledge. This insight is crucial to all of modern science. The old paradigm is based on the Cartesian belief in the certainty of scientific knowledge. In the new paradigm it is recognized that all scientific concepts and theories are limited and approximate. Science can never provide any complete and definitive understanding.

This can be illustrated easily with a simple experiment that is often performed in introductory physics courses. The professor drops an object from a certain height and shows her students with a simple formula from Newtonian physics how to calculate the time it takes for the object to reach the ground. As with most of Newtonian physics, this calculation will neglect the resistance of the air and will therefore not be completely accurate. Indeed, if the object to be dropped were a feather, the experiment would not work at all.

The professor may be satisfied with this "first approximation," or she may want to go a step further and take the air resistance into account by adding a simple term to the formula. The result—the second approximation—will be more accurate but still not completely so, because air resistance depends on the temperature and pressure of the air. If the professor is very ambitious, she may derive a much more complicated formula as a third approximation, which would take these variables into account.

However, the air resistance depends not only on the temperature and air pressure, but also on the air convection—that is, on the large-scale circulation of air particles through the room. The students may observe that this air convection is caused, in addition to an open window, by their breathing patterns; and at this point the professor will probably stop the process of improving the approximation in successive steps.

This simple example shows that the fall of an object is connected in multiple ways to its environment—and, ultimately, to the rest of the universe. No matter how many connections we take into account in our scientific description of a phenomenon, we will always be forced to leave others out. Therefore scientists can never deal with truth, in the sense of a precise correspondence between the description and the described phenomenon. In science we always deal with limited and approximate descriptions of reality. This may sound frustrating, but for systems thinkers the fact that we *can* obtain approximate knowledge about an infinite web of interconnected patterns is a source of confidence and strength. Louis Pasteur said it beautifully:

> Science advances through tentative answers to a series of more and more subtle questions which reach deeper and deeper into the essence of natural phenomena.[7]

Process Thinking

All the systems concepts discussed so far can be seen as different aspects of one great strand of systemic thinking, which we may call contextual thinking. There is another strand of equal importance, which emerged somewhat later in twentieth-century science. This second strand is process thinking. In the mechanistic framework of Cartesian science there are fundamental structures, and then there are forces and mechanisms through which these interact, thus giving rise to processes. In systems science every structure is seen as the manifestation of underlying processes. Systems thinking is always process thinking.

In the development of systems thinking during the first half of the century, the process aspect was first emphasized by the Austrian biologist Ludwig von Bertalanffy in the late 1930s and was further explored in cybernetics during the 1940s. Once the cyberneticists had made feedback loops and other dynamic patterns a central subject of scientific investigation, ecologists began to study the cyclical flows of matter and energy through ecosystems. For example, Eugene Odum's text *Fundamentals of Ecology,* which influenced a whole generation of ecologists, depicted ecosystems in terms of simple flow diagrams.[8]

Of course, like contextual thinking, process thinking, too, had its forerunners, even in Greek antiquity. Indeed, at the dawn of Western science we encounter Heraclitus' celebrated dictum: "Everything flows." During the 1920s the English mathematician and philosopher Alfred North Whitehead formulated a strongly process-oriented philosophy.[9] At the same time the physiologist Walter Cannon took up Claude Bernard's principle of the constancy of an organism's "internal environment" and refined it into the concept of homeostasis—the self-regulatory mechanism that allows organisms to maintain themselves in a state of dynamic balance with their variables fluctuating between tolerance limits.[10]

In the meantime, detailed experimental studies of cells had made it clear that the metabolism of a living cell combines order and activity in a way that cannot be described by mechanistic science. It involves thousands of chemical reactions, all taking place simultaneously to transform the cell's nutrients, synthesize its basic structures, and eliminate its waste products. Metabolism is a continual, complex, and highly organized activity.

Whitehead's process philosophy, Cannon's concept of homeostasis, and the experimental work on metabolism all had a strong influence on Ludwig von Bertalanffy, leading him to formulate a new theory of "open systems." Later on, during the 1940s, Bertalanffy enlarged his framework and attempted to combine the various concepts of systems thinking and organismic biology into a formal theory of living systems.

Tektology

Ludwig von Bertalanffy is commonly credited with the first formulation of a comprehensive theoretical framework describing the principles of organization of living systems. However, twenty to thirty years before he published the first papers on his "general systems theory," Alexander Bogdanov, a Russian medical researcher, philosopher, and economist, developed a systems theory of equal sophistication and scope, which unfortunately is still largely unknown outside of Russia.[11]

Bogdanov called his theory "tektology," from the Greek *tekton* ("builder"), which can be translated as "the science of structures." Bogdanov's main goal was to clarify and generalize the principles of organization of all living and nonliving structures:

> Tektology must clarify the modes of organization that are perceived to exist in nature and human activity; then it must generalize and systematize these modes; further it must explain them, that is, propose abstract schemes of their tendencies and laws. . . . Tektology deals with organizational experiences not of this or that specialized field, but of all these fields together. In other words, tektology embraces the subject matter of all the other sciences.[12]

Tektology was the first attempt in the history of science to arrive at a systematic formulation of the principles of organization operating in living and nonliving systems.[13] It anticipated the conceptual framework of Ludwig von Bertalanffy's general systems theory, and it also included several important ideas that were formulated four decades later, in a different language, as key principles of cybernetics by Norbert Wiener and Ross Ashby.[14]

Bogdanov's goal was to formulate a "universal science of organization." He defined organizational form as "the totality of connections among systemic elements," which is virtually identical to our contemporary definition of pattern of organization.[15] Using the terms "complex" and "system" interchangeably, Bogdanov distinguished three kinds of systems: organized complexes, where the whole is greater than the sum of its parts; disorganized complexes, where the whole is smaller than the sum of its parts; and neutral complexes, where the organizing and disorganizing activities cancel each other.

The stability and development of all systems can be understood, according to Bogdanov, in terms of two basic organizational mechanisms: formation and regulation. By studying both forms of organizational dynamics and illustrating them with numerous examples from natural and social systems, Bogdanov explores several key ideas pursued by organismic biologists *and* by cyberneticists.

The dynamics of formation consists in the joining of complexes through various kinds of linkages, which Bogdanov analyzes in great detail. He emphasizes in particular that the tension between crisis and transformation is central to the formation of complex systems. Foreshadowing the work of Ilya Prigogine,[16] Bogdanov shows how organizational crisis manifests itself as a breakdown of the existing systemic balance and at the same time represents an organizational transition to a new state of balance. By defining categories of crises, Bogdanov even anticipates the concept of catastrophe developed by the French mathematician René Thom, which is a key ingredient in the currently emerging new mathematics of complexity.[17]

Like Bertalanffy, Bogdanov recognized that living systems are open systems that operate far from equilibrium, and he carefully studied their regulation and self-regulation processes. A system for which there is no need of external regulation, because the system regulates itself, is called "bi-regulator" in Bogdanov's language. Using the example of the steam engine to illustrate self-regulation, as the cyberneticists would do several decades later, Bogdanov essentially described the mechanism defined as feedback by Norbert Wiener, which became a central concept of cybernetics.[18]

Bogdanov did not attempt to formulate his ideas mathematically, but he did envisage the future development of an abstract "tektological symbolism," a new kind of mathematics to analyze the patterns of organization he had discovered. Half a century later such a new mathematics has indeed emerged.[19]

Bogdanov's pioneering book, *Tektology,* was published in Russian in three volumes between 1912 and 1917. A German edition was published and widely reviewed in 1928. However, very little is known in the West about this first version of a general systems theory and precursor of cybernetics. Even in Ludwig von Bertalanffy's *General System Theory,* published in 1968, which includes a section on the history of systems theory, there is no reference to Bogdanov whatsoever. It is difficult to understand how Bertalanffy, who was widely read and published all his original work in German, would not have come across Bogdanov's work.[20]

Among his contemporaries Bogdanov was largely misunderstood because he was so far ahead of his time. In the words of the Azerbaijani scientist A. L. Takhtadzhian: "Foreign in its universality to the scientific thinking of the time, the idea of a general theory of organization was fully understood only by a handful of men and did not therefore spread."[21]

Marxist philosophers of the day were hostile to Bogdanov's ideas because they perceived tektology as a new philosophical system designed to replace that of Marx, even though Bogdanov protested repeatedly against the confusion of his universal science of organization with philosophy. Lenin mercilessly attacked Bogdanov as a philosopher, and consequently his works were suppressed for almost half a century in the Soviet Union. Recently, however, in the wake of Gorbachev's perestroika, Bogdanov's writings have received great attention from Russian scientists and philosophers. Thus it is to be hoped that Bogdanov's pioneering work will now be recognized more widely also outside Russia.

General Systems Theory

Before the 1940s the terms "system" and "systems thinking" had been used by several scientists, but it was Bertalanffy's concepts of an open system and a general systems theory that established systems thinking as a major scientific movement.[22] With the subsequent strong support from cybernetics, the concepts of systems thinking and systems theory became integral parts of the established scientific language and led to numerous new methodologies and applications—systems engineering, systems analysis, systems dynamics, and so on.[23]

Ludwig von Bertalanffy began his career as a biologist in Vienna during the 1920s. He soon joined a group of scientists and philosophers, known internationally as the Vienna Circle, and his work included broader philosophical themes from the very beginning.[24] Like other organismic biologists, he firmly believed that biological phenomena required new ways of thinking, transcending the traditional methods of the physical sciences. He set out to replace the mechanistic foundations of science with a holistic vision:

> General system theory is a general science of "wholeness" which up till now was considered a vague, hazy, and semi-metaphysical concept. In elaborate form it would be a mathematical discipline, in itself purely formal but applicable to the various empirical sciences. For sciences concerned with "organized wholes," it would be of similar significance to that which probability theory has for sciences concerned with "chance events."[25]

In spite of this vision of a future formal, mathematical theory, Bertalanffy sought to establish his general systems theory on a solid biological basis. He objected to the dominant position of physics within modern science and emphasized the crucial difference between physical and biological systems.

To make his point, Bertalanffy pinpointed a dilemma that had puzzled scientists since the nineteenth century, when the novel idea of evolution entered into scientific thinking. Whereas Newtonian mechanics was a science of forces and trajectories, evolutionary thinking—thinking in terms of change, growth, and development—required a new science of complexity.[26] The first formulation of this new science was classical thermodynamics with its celebrated "second law," the law of the dissipation of energy.[27] According to the second law of thermodynamics, formulated first by the French physicist Sadi Carnot in terms of the technology of thermal engines, there is a trend in physical phenomena from order to disorder. Any isolated, or "closed," physical system will proceed spontaneously in the direction of ever-increasing disorder.

To express this direction in the evolution of physical systems in precise mathematical form, physicists introduced a new quantity called "entropy."[28] According to the second law, the entropy of a closed physical system will keep increasing, and because this evolution is accompanied by increasing disorder, entropy can also be seen as a measure of disorder.

With the concept of entropy and the formulation of the second law, thermodynamics introduced the idea of irreversible processes, of an "arrow of time," into science. According to the second law, some mechanical energy is always dissipated into heat that cannot be completely recovered. Thus the entire world machine is running down and will eventually grind to a halt.

This grim picture of cosmic evolution was in sharp contrast with the evolutionary thinking among nineteenth-century biologists, who observed that the living universe evolves from disorder to order, toward states of ever-increasing complexity. At the end of the nineteenth century, then, Newtonian mechanics, the science of eternal, reversible trajectories, had been supplemented by two diametrically opposed views of evolutionary change—that of a living world unfolding toward increasing order and complexity and that of an engine running down, a world of ever-increasing disorder. Who was right, Darwin or Carnot?

Ludwig von Bertalanffy could not resolve this dilemma, but he took the crucial first step by recognizing that living organisms are open systems that cannot be described by classical thermodynamics. He called such systems "open" because they need to feed on a continual flux of matter and energy from their environment to stay alive:

> The organism is not a static system closed to the outside and always containing the identical components; it is an open system in a (quasi-) steady state . . . in which material continually enters from, and leaves into, the outside environment.[29]

Unlike closed systems, which settle into a state of thermal equilibrium, open systems maintain themselves far from equilibrium in this "steady state" characterized by continual flow and change. Bertalanffy coined the German term *Fliessgleichgewicht* ("flowing balance") to describe such a state of dynamic balance. He recognized clearly that classical thermodynamics, which deals with closed systems at or near equilibrium, is inappropriate to describe open systems in steady states far from equilibrium.

In open systems, Bertalanffy speculated, entropy (or disorder) may decrease, and the second law of thermodynamics may not apply. He postulated that classical science would have to be complemented by a new thermodynamics of open systems. However, in the 1940s the mathematical techniques required for such an expansion of thermodynamics were not available to Bertalanffy. The formulation of the new thermodynamics of open systems had to wait until the 1970s. It was the great achievement of Ilya Prigogine, who used a new mathematics to reevaluate the second law by radically rethinking traditional scientific views of order and disorder, which enabled him to resolve unambiguously the two contradictory nineteenth-century views of evolution.[30]

Bertalanffy correctly identified the characteristics of the steady state as those of the process of metabolism, which led him to postulate self-regulation as another key property of open systems. This idea was refined by Prigogine thirty years later in terms of the self-organization of "dissipative structures."[31]

Ludwig von Bertalanffy's vision of a "general science of wholeness" was based on his observation that systemic concepts and principles can be applied in many different fields of study: "The parallelism of general conceptions or even special laws in different fields," he explained, "is a consequence of the fact that these are concerned with 'systems,' and that certain general principles apply to systems irrespective of their nature."[32] Since living systems span such a wide range of phenomena, involving individual organisms and their parts, social systems, and ecosystems, Bertalanffy believed that a general systems theory would offer an ideal conceptual framework for unifying various scientific disciplines that had become isolated and fragmented:

> General system theory should be . . . an important means of controlling and instigating the transfer of principles from one field to another, and it will no longer be necessary to duplicate or triplicate the discovery of the same principle in different fields isolated from each other. At the same time, by formulating exact criteria, general system theory will guard against superficial analogies which are useless in science.[33]

Bertalanffy did not see the realization of his vision, and a general science of wholeness of the kind he envisaged may never be formulated. However, during the two decades after his death in

1972, a systemic conception of life, mind, and consciousness began to emerge that transcends disciplinary boundaries and, indeed, holds the promise of unifying various fields of study that were formerly separated. Although this new conception of life has its roots more clearly in cybernetics than in general systems theory, it certainly owes a great deal to the concepts and thinking that Ludwig von Bertalanffy introduced into science.

4

The Logic of the Mind

While Ludwig von Bertalanffy worked on his general systems theory, attempts to develop self-guiding and self-regulating machines led to an entirely new field of investigation that had a major impact on the further development of the systems view of life. Drawing from several disciplines, the new science represented a unified approach to problems of communication and control, involving a whole complex of novel ideas, which inspired Norbert Wiener to invent a special name for it—"cybernetics." The word is derived from the Greek *kybernetes* ("steersman"), and Wiener defined cybernetics as the science of "control and communication in the animal and the machine."[1]

The Cyberneticists

Cybernetics soon became a powerful intellectual movement, which developed independently of organismic biology and general systems theory. The cyberneticists were neither biologists nor ecologists; they were mathematicians, neuroscientists, social scientists, and engineers. They were concerned with a different level of description, concentrating on patterns of communication, especially in closed loops and networks. Their investigations led them to the concepts of feedback and self-regulation and then, later on, to self-organization.

This attention to patterns of organization, which was implicit in organismic biology and Gestalt psychology, became the explicit focus of cybernetics. Wiener, especially, recognized that the new notions of message, control, and feedback referred to patterns of organization—that is, to nonmaterial entities—that are crucial to a full scientific description of life. Later on Wiener expanded the concept of pattern, from the patterns of communication and control that are common to animals and machines to the general idea of pattern as a key characteristic of life. "We are but whirlpools in a river of ever-flowing water," he wrote in 1950. "We are not stuff that abides, but patterns that perpetuate themselves."[2]

The cybernetics movement began during World War II, when a group of mathematicians, neuroscientists, and engineers—among them Norbert Wiener, John von Neumann, Claude Shannon, and Warren McCulloch—formed an informal network to pursue common scientific interests.[3] Their work was closely linked to military research that dealt with the problems of tracking and shooting down aircraft and was funded by the military, as was most subsequent research in cybernetics.

The first cyberneticists (as they would call themselves several years later) set themselves the challenge of discovering the neural mechanisms underlying mental phenomena and expressing them in explicit mathematical language. Thus while the organismic biologists were concerned with the material side of the Cartesian split, revolting against mechanism and exploring the nature of biological form, the cyberneticists turned to the mental side. Their intention from the beginning was to create an exact science of mind.[4] Although their approach was quite mechanistic, concentrating on patterns common to animals and machines, it involved many novel ideas that exerted a tremendous influence on subsequent systemic conceptions of mental phenomena. Indeed, the contemporary science of cognition, which offers a unified scientific conception of brain and mind, can be traced back directly to the pioneering years of cybernetics.

The conceptual framework of cybernetics was developed in a

series of legendary meetings in New York City, known as the Macy Conferences.[5] These meetings—especially the first one in 1946—were extremely stimulating, bringing together a unique group of highly creative people who engaged in intense interdisciplinary dialogues to explore new ideas and ways of thinking. The participants fell into two core groups. The first formed around the original cyberneticists and consisted of mathematicians, engineers, and neuroscientists. The other group consisted of scientists from the humanities who clustered around Gregory Bateson and Margaret Mead. From the first meeting on, the cyberneticists made great efforts to bridge the academic gap between themselves and the humanities.

Norbert Wiener was the dominant figure throughout the conference series, imbuing it with his enthusiasm for science and dazzling his fellow participants with the brilliance of his ideas and often irreverent approaches. According to many witnesses Wiener had the disconcerting tendency to fall asleep during discussions, and even to snore, apparently without losing track of what was being said. Upon waking up, he would immediately make detailed and penetrating comments or point out logical inconsistencies. He thoroughly enjoyed these discussions and his central role in them.

Wiener was not only a brilliant mathematician, he was also an articulate philosopher. (In fact, his degree from Harvard was in philosophy.) He was keenly interested in biology and appreciated the richness of natural, living systems. He looked beyond the mechanisms of communication and control to larger patterns of organization and tried to relate his ideas to a wide range of social and cultural issues.

John von Neumann was the second center of attraction at the Macy Conferences. A mathematical genius, he had written a classic treatise on quantum theory, was the originator of the theory of games, and became world famous as the inventor of the digital computer. Von Neumann had a powerful memory, and his mind worked with enormous speed. It was said of him that he could understand the essence of a mathematical problem almost instantly and that he would analyze any problem, mathematical or practical, so clearly and exhaustively that no further discussion was necessary.

At the Macy meetings von Neumann was fascinated by the processes of the human brain and saw the description of brain functioning in formal logical terms as the ultimate challenge of science. He had tremendous confidence in the power of logic and great faith in technology, and throughout his work he looked for universal logical structures of scientific knowledge.

Von Neumann and Wiener had much in common.[6] Both were admired as mathematical geniuses, and their influence on society was far stronger than that of other mathematicians of their generation. They both trusted their subconscious minds. Like many poets and artists, they had the habit of sleeping with pencil and paper near their beds and made use of the imagery of their dreams in their work. However, these two pioneers of cybernetics differed significantly in their approach to science. Whereas von Neumann looked for control, for a program, Wiener appreciated the richness of natural patterns and sought a comprehensive conceptual synthesis.

In keeping with these characteristics, Wiener stayed away from people with political power, whereas von Neumann felt very comfortable in their company. At the Macy Conferences their different attitudes toward power, and especially toward military power, was the source of growing friction, which eventually led to a complete break. Whereas von Neumann remained a military consultant throughout his career, specializing in the application of computers to weapons systems, Wiener ended his military work shortly after the first Macy meeting. "I do not expect to publish any future work of mine," he wrote at the end of 1946, "which may do damage in the hands of irresponsible militarists."[7]

Norbert Wiener had a strong influence on Gregory Bateson, with whom he had a very good rapport throughout the Macy Conferences. Bateson's mind, like Wiener's, roamed freely across disciplines, challenging the basic assumptions and methods of several sciences by searching for general patterns and powerful universal abstractions. Bateson thought of himself primarily as a biologist and considered the many fields he became involved in—

anthropology, epistemology, psychiatry, and others—as branches of biology. The great passion he brought to science embraced the full diversity of phenomena associated with life, and his main aim was to discover common principles of organization in that diversity—"the pattern which connects," as he would put it many years later.[8] At the cybernetics conferences Bateson and Wiener both searched for comprehensive, holistic descriptions while being careful to remain within the boundaries of science. In so doing, they created a systems approach to a broad range of phenomena.

His dialogues with Wiener and the other cyberneticists had a lasting impact on Bateson's subsequent work. He pioneered the application of systems thinking to family therapy, developed a cybernetic model of alcoholism, and authored the double-bind theory of schizophrenia, which had a major impact on the work of R. D. Laing and many other psychiatrists. However, Bateson's most important contribution to science and philosophy may have been the concept of mind, based on cybernetic principles, which he developed during the 1960s. This revolutionary work opened the door to understanding the nature of mind as a systems phenomenon and became the first successful attempt in science to overcome the Cartesian division between mind and body.[9]

The series of ten Macy Conferences was chaired by Warren McCulloch, professor of psychiatry and physiology at the University of Illinois, who had a solid reputation in brain research and made sure that the challenge of reaching a new understanding of mind and brain remained at the center of the dialogues.

The pioneering years of cybernetics resulted in an impressive series of concrete achievements, in addition to the lasting impact on systems thinking as a whole, and it is amazing that most of the novel ideas and theories were discussed, at least in their outlines, at the very first meeting.[10] The first conference began with an extensive description of digital computers (which had not yet been built) by John von Neumann, followed by von Neumann's persuasive presentation of analogies between the computer and the brain. The basis of these analogies, which were to dominate the cyberneticists' view of cognition for the subsequent three decades, was the use of mathematical logic to understand brain functioning, one of the outstanding achievements of cybernetics.

Von Neumann's presentations were followed by Norbert Wiener's detailed discussion of the central idea of his work, the concept of feedback. Wiener then introduced a cluster of new ideas, which coalesced over the years into information theory and communication theory. Gregory Bateson and Margaret Mead concluded the presentations with a review of the conceptual framework of the social sciences, which they considered inadequate and in need of basic theoretical work inspired by the new cybernetic concepts.

Feedback

All the major achievements of cybernetics originated in comparisons between organisms and machines—in other words, in mech-

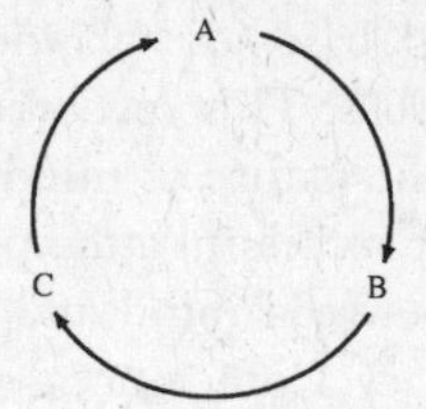

Figure 4-1
Circular causality of a feedback loop.

anistic models of living systems. However, the cybernetic machines are very different from Descartes's clockworks. The crucial difference is embodied in Norbert Wiener's concept of feedback and is expressed in the very meaning of "cybernetics." A feedback loop is a circular arrangement of causally connected elements, in which an initial cause propagates around the links of the loop, so that each element has an effect on the next, until the last "feeds back" the effect into the first element of the cycle (see figure 4-1). The consequence of this arrangement is that the first link ("input") is affected by the last ("output"), which results in self-regulation of the entire system, as the initial effect is modified each time it travels around the cycle. Feedback, in Wiener's words, is the "control of a machine on the basis of its *actual* performance rather than its *expected* performance."[11] In a broader sense feedback has come to mean the conveying of information about the outcome of any process or activity to its source.

Wiener's original example of the steersman is one of the simplest examples of a feedback loop (see figure 4-2). When the boat deviates from the preset course—say, to the right—the steersman assesses the deviation and then countersteers by moving the rudder to the left. This decreases the boat's deviation, perhaps even to the point of moving through the correct position and then deviating to the left. At some time during this movement the steersman makes a new assessment of the boat's deviation, countersteers accordingly, assesses the deviation again, and so on. Thus he relies on continual feedback to keep the boat on course, its actual trajectory oscillating around the preset direction. The skill of steering a boat consists in keeping these oscillations as smooth as possible.

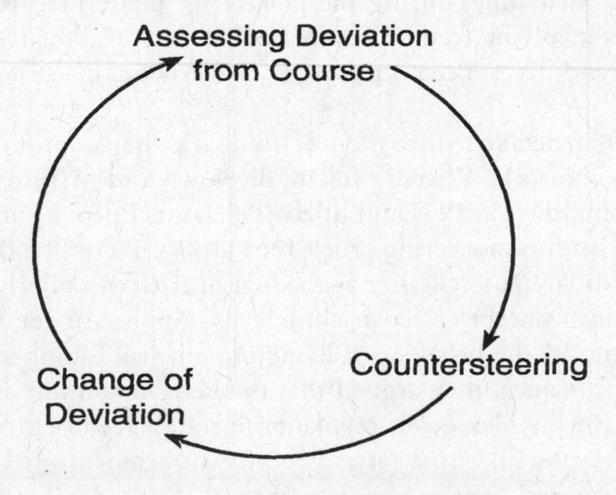

Figure 4-2
Feedback loop representing the steering of a boat.

A similar feedback mechanism is in play when we ride a bicycle. At first, when we learn to do so, we find it difficult to monitor the feedback from the continual changes of balance and to steer the bicycle accordingly. Thus a beginner's front wheel tends to oscillate strongly. But as our expertise increases, our brain monitors, evaluates, and responds to the feedback automatically, and the oscillations of the front wheel smooth out into a straight line.

Self-regulating machines involving feedback loops existed long before cybernetics. The centrifugal governor of a steam engine, invented by James Watt in the late eighteenth century, is a classic example, and the first thermostats were invented even earlier.[12] The engineers who designed these early feedback devices described their operations and pictured their mechanical components in design sketches, but they never recognized the pattern of circular causality embedded in them. In the nineteenth century the famous physicist James Clerk Maxwell wrote a formal mathematical analysis of the steam governor without ever mentioning the underlying loop concept. Another century had to go by before the connection between feedback and circular causality was recognized. At that time, during the pioneering phase of cybernetics, machines involving feedback loops became a central focus of engineering and have been known as "cybernetic machines" ever since.

The first detailed discussion of feedback loops appeared in a paper by Norbert Wiener, Julian Bigelow, and Arturo Rosenblueth, published in 1943 and titled "Behavior, Purpose, and Teleology."[13] In this pioneering article the authors not only introduced the idea of circular causality as the logical pattern underlying the engineering concept of feedback, but also applied it for the first time to model the behavior of living organisms. Taking a strictly behaviorist stance, they argued that the behavior of any machine or organism involving self-regulation through feedback could be called "purposeful," since it is behavior directed toward a goal. They illustrated their model of such goal-directed behavior with numerous examples—a cat catching a mouse, a dog following a trail, a person lifting a glass from a table, and so on—analyzing them in terms of the underlying circular feedback patterns.

Wiener and his colleagues also recognized feedback as the essential mechanism of homeostasis, the self-regulation that allows living organisms to maintain themselves in a state of dynamic balance. When Walter Cannon introduced the concept of homeostasis a decade earlier in his influential book *The Wisdom of the Body,*[14] he gave detailed descriptions of many self-regulatory metabolic processes but never explicitly identified the closed causal loops embodied in them. Thus the concept of the feedback loop introduced by the cyberneticists led to new perceptions of the many self-regulatory processes characteristic of life. Today we understand that feedback loops are ubiquitous in the living world, because they are a special feature of the nonlinear network patterns that are characteristic of living systems.

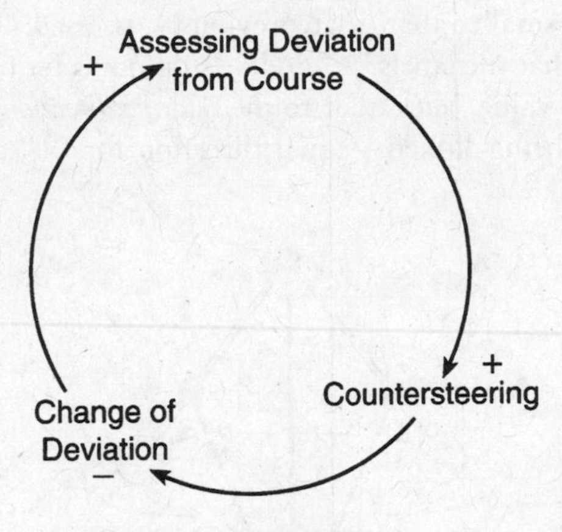

Figure 4-3
Positive and negative causal links.

The cyberneticists distinguished between two kinds of feedback—self-balancing (or "negative") and self-reinforcing (or "positive") feedback. Examples of the latter are the commonly known runaway effects, or vicious circles, in which the initial effect continues to be amplified as it travels repeatedly around the loop.

Since the technical meanings of "negative" and "positive" in this context can easily give rise to confusion, it may be worthwhile to explain them in more detail.[15] A causal influence from A to B is defined as positive if a change in A produces a change in B in the same direction—for example, an increase of B if A increases and a decrease if A decreases. The causal link is defined as negative if B changes in the opposite direction, decreasing if A increases and increasing if A decreases.

For example, in the feedback loop representing the steering of a boat, redrawn in figure 4-3, the link between "assessing deviation" and "countersteering" is positive—the greater the deviation from the preset course, the greater the amount of countersteering. The next link, however, is negative—the more the countersteering increases, the sharper the deviation will decrease. Finally, the last link is again positive. As the deviation decreases, its newly assessed value will be smaller than that previously assessed. The point to remember is that the labels "+" and "–" do not refer to an increase or decrease of value, but rather to the *relative direction of change* of the elements being linked—equal direction for "+" and opposite direction for "–".

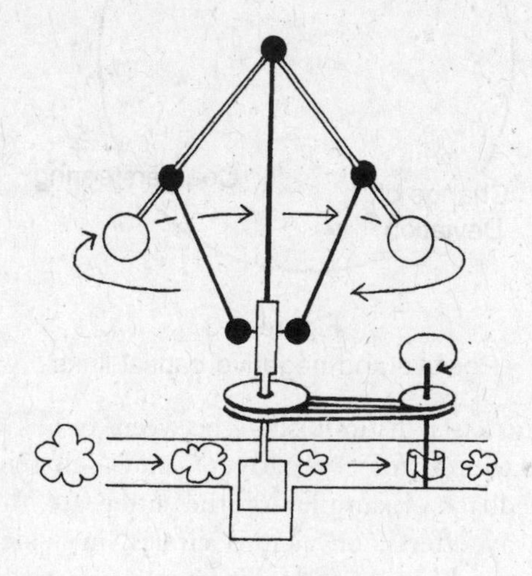

Figure 4-4
Centrifugal governor.

The reason why these labels are so convenient is that they lead to a very simple rule for determining the overall character of the feedback loop. It will be self-balancing ("negative") if it contains an odd number of negative links and self-reinforcing ("positive") if it contains an even number of negative links.[16] In our example there is only one negative link; so the entire loop is negative, or self-balancing. Feedback loops are frequently composed of both positive and negative causal links, and their overall character is easily determined simply by counting the number of negative links around the loop.

The examples of steering a boat and riding a bicycle are ideally suited to illustrate the feedback concept, because they refer to well-known human experiences and are thus understood immediately. To illustrate the same principles with a mechanical device for self-regulation, Wiener and his colleagues often used one of the earliest and simplest examples of feedback engineering, the centrifugal governor of a steam engine (see figure 4-4). It consists of a rotating spindle with two weights ("flyballs") attached to it in such a way that they move apart, driven by the centrifugal force, when the speed of the rotation increases. The governor sits on top of the steam engine's cylinder, and the weights are connected with a piston, which cuts off the steam as they move apart. The pressure of the steam drives the engine, which drives a flywheel. The flywheel, in turn, drives the governor, and thus the loop of cause and effect is closed.

The feedback sequence is easily read off from the loop diagram drawn in figure 4-5. An increase in the speed of the engine increases the rotation of the governor. This increases the distance between the weights, which cuts down the steam supply. As the steam supply decreases, the speed of the engine decreases as well; the rotation of the governor slows down; the weights move closer together; steam supply increases; the engine speeds up again; and so on. The only negative link in the loop is the one between "distance between weights" and "steam supply," and therefore the entire feedback loop is negative, or self-balancing.

From the beginning of cybernetics, Norbert Wiener was aware that feedback is an important concept for modeling not only living organisms but also social systems. Thus he wrote in *Cybernetics:*

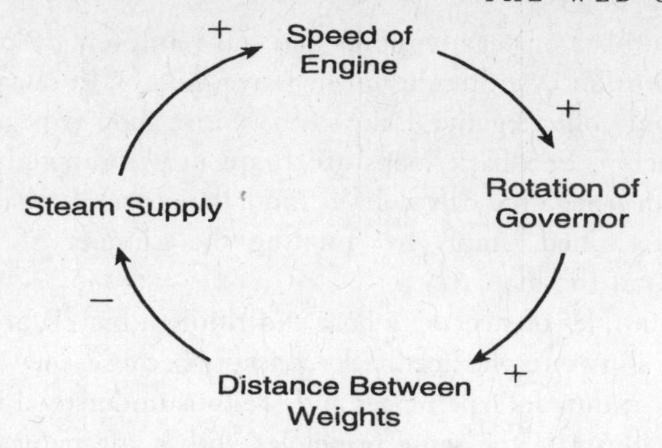

Figure 4-5
Feedback loop for centrifugal governor.

> It is certainly true that the social system is an organization like the individual, that is bound together by a system of communication, and that it has a dynamics in which circular processes of a feedback nature play an important role.[17]

It was the discovery of feedback as a general pattern of life, applicable to organisms and social systems, which got Gregory Bateson and Margaret Mead so excited about cybernetics. As social scientists they had observed many examples of circular causality implicit in social phenomena, and during the Macy meetings the dynamics of these phenomena were made explicit in a coherent unifying pattern.

Throughout the history of the social sciences numerous metaphors have been used to describe self-regulatory processes in social life. The best known, perhaps, are the "invisible hand" regulating the market in the economic theory of Adam Smith, the "checks and balances" of the U.S. Constitution, and the interplay of thesis and antithesis in the dialectic of Hegel and Marx. The phenomena described by these models and metaphors all imply circular patterns of causality that can be represented by feedback loops, but none of their authors made that fact explicit.[18]

If the circular logical pattern of self-balancing feedback was not recognized before cybernetics, that of self-reinforcing feedback had been known for hundreds of years in common parlance as a "vicious circle." The expressive metaphor describes a bad situation leading to its own worsening through a circular sequence of events. Perhaps the circular nature of such self-reinforcing, "runaway" feedback loops was recognized explicitly much earlier, because their effect is much more dramatic than the self-balancing of the negative feedback loops that are so widespread in the living world.

There are other common metaphors to describe self-reinforcing feedback phenomena.[19] The "self-fulfilling prophecy," in which originally unfounded fears lead to actions that make the fears come true, and the "bandwagon effect"—the tendency of a cause to gain support simply because of its growing number of adherents—are two well-known examples.

In spite of the extensive knowledge of self-reinforcing feedback in common folk wisdom, it played hardly any role during the first phase of cybernetics. The cyberneticists around Norbert Wiener acknowledged the existence of runaway feedback phenomena but did not study them any further. Instead they concentrated on the self-regulatory, homeostatic processes in living organisms. Indeed, purely self-reinforcing feedback phenomena are rare in nature, as they are usually balanced by negative feedback loops constraining their runaway tendencies.

In an ecosystem, for example, every species has the potential of undergoing an exponential population growth, but these tendencies are kept in check by various balancing interactions within the system. Exponential runaways will appear only when the ecosystem is severely disturbed. Then some plants will turn into "weeds," some animals become "pests," and other species will be exterminated, and thus the balance of the whole system will be threatened.

During the 1960s anthropologist and cyberneticist Magoroh Maruyama took up the study of self-reinforcing, or "deviation-amplifying" feedback processes in a widely read article, titled "The Second Cybernetics."[20] He introduced the feedback diagrams with "+" and "–" labels attached to their causal links, and he used this convenient notation for a detailed analysis of the interplay of negative and positive feedback processes in biological and social phenomena. In doing so, he linked the feedback concept of cybernetics with the notion of "mutual causality," which had been developed by social scientists in the meantime, and thus contributed significantly to the influence of cybernetic principles on social thought.[21]

From the point of view of the history of systems thinking, one of the most important aspects of the cyberneticists' extensive studies of feedback loops is the recognition that they depict patterns of organization. The circular causality in a feedback loop does not imply that the elements in the corresponding physical system are arranged in a circle. Feedback loops are abstract patterns of relationships embedded in physical structures or in the activities of living organisms. For the first time in the history of systems thinking, the cyberneticists clearly distinguished the pattern of organization of a system from its physical structure—a distinction that is crucial in the contemporary theory of living systems.[22]

Information Theory

An important part of cybernetics was the theory of information developed by Norbert Wiener and Claude Shannon in the late 1940s. It originated in Shannon's attempts at the Bell Telephone Laboratories to define and measure amounts of information transmitted through telegraph and telephone lines in order to estimate efficiencies and establish a basis for charging for messages.

The term "information" is used in information theory in a highly technical sense, which is quite different from our everyday use of the word and has nothing to do with meaning. This has resulted in endless confusion. According to Heinz von Foerster, a regular participant in the Macy Conferences and editor of the written proceedings, the whole problem is based on a very unfortunate linguistic error—the confusion between "information" and "signal," which led the cyberneticists to call their theory a theory of information rather than a theory of signals.[23]

Information theory, then, is concerned mainly with the problem of how to get a message, coded as a signal, through a noisy channel. However, Norbert Wiener also emphasized the fact that such a coded message is essentially a pattern of organization, and by drawing an analogy between such patterns of communication and the patterns of organization in organisms, he further prepared the ground for thinking about living systems in terms of patterns.

Cybernetics of the Brain

During the 1950s and 1960s Ross Ashby became the leading theorist of the cybernetics movement. Like McCulloch, Ashby was a neurologist by training, but he went much further than McCulloch in exploring the nervous system and constructing cybernetic models of neural processes. In his book *Design for a Brain,* Ashby attempted to explain in purely mechanistic and deterministic terms the brain's unique adaptive behavior, capacity for memory, and other patterns of brain functioning. "It will be assumed," he wrote, "that a machine or an animal behaved in a certain way at a certain moment because its physical and chemical nature at that moment allowed no other action."[24]

It is evident that Ashby was much more Cartesian in his approach to cybernetics than Norbert Wiener, who made a clear distinction between a mechanistic model and the nonmechanistic living system it represents. "When I compare the living organism with . . . a machine," wrote Wiener, "I do not for a moment mean that the specific physical, chemical, and spiritual processes of life as we ordinarily know it are the same as those of life-imitating machines."[25]

In spite of his strictly mechanistic outlook, Ross Ashby advanced the fledgling discipline of cognitive science considerably with his detailed analyses of sophisticated cybernetic models of neural processes. In particular he clearly recognized that living systems are energetically open while being—in today's terminology—organizationally closed: "Cybernetics might . . . be defined," wrote Ashby, "as the study of systems that are open to energy but closed to information and control—systems that are 'information-tight.' "[26]

Computer Model of Cognition

When the cyberneticists explored patterns of communication and control, the challenge to understand "the logic of the mind" and express it in mathematical language was always at the very center of their discussions. Thus for over a decade the key ideas of cybernetics were developed through a fascinating interplay among biology, mathematics, and engineering. Detailed studies of the human nervous system led to the model of the brain as a logical circuit with neurons as its basic elements. This view was crucial for the invention of digital computers, and that technological breakthrough in turn provided the conceptual basis for a new approach to the scientific study of mind. John von Neumann's invention of the computer and his analogy between computer and brain functioning are so closely intertwined that it is difficult to know which came first.

The computer model of mental activity became the prevalent view of cognitive science and dominated all brain research for the next thirty years. The basic idea was that human intelligence resembles that of a computer to such an extent that cognition—the process of knowing—can be defined as information processing—in other words, as manipulation of symbols based on a set of rules.[27]

The field of artificial intelligence developed as a direct consequence of this view, and soon the literature was full of outrageous claims about computer "intelligence." Thus Herbert Simon and Allen Newell wrote as early as 1958:

> There are now in the world machines that think, that learn and that create. Moreover, their ability to do these things is going to increase rapidly until—in the visible future—the range of problems they can handle will be coextensive with the range to which the human mind has been applied.[28]

This prediction is as absurd today as it was thirty-eight years ago, yet it is still widely believed. The enthusiasm among scientists and the general public for the computer as a metaphor for the human brain has an interesting parallel in the enthusiasm of Descartes and his contemporaries for the clock as a metaphor for the body.[29] For Descartes the clock was a unique machine. It was the only machine that functioned autonomously, running by itself once it was wound up. This was the time of the French Baroque, when clock mechanisms were widely used to build artful "lifelike" machinery, which delighted people with the magic of their seemingly spontaneous movements. Like most of his contemporaries, Descartes was fascinated by these automata, and he found it natural to compare their functioning to that of living organisms:

> We see clocks, artificial fountains, mills and other similar machines which, though merely man-made, have nonetheless the power to move by themselves in several different ways. . . . I do not recognize any difference between the machines made by craftsmen and the various bodies that nature alone composes.[30]

The clockworks of the seventeenth century were the first autonomous machines, and for three hundred years they were the only machines of their kind—until the invention of the computer. The computer is again a novel and unique machine. It not only moves autonomously once it is programmed and turned on, it does something completely new: it processes information. And since von Neumann and the early cyberneticists believed that the human brain, too, processes information, it was natural for them to use the computer as a metaphor for the brain and even for the mind, just as it had been for Descartes to use the clock as a metaphor for the body.

Like the Cartesian model of the body as a clockwork, that of the brain as a computer was very useful at first, providing an exciting framework for a new scientific understanding of cognition and leading to many fresh avenues of research. By the mid-1960s, however, the original model, which encouraged the exploration of its own limitations and the discussion of alternatives, had hardened into a dogma, as so often happens in science. During the subsequent decade almost all of neurobiology was dominated by the information-processing perspective, whose origins and underlying assumptions were hardly even questioned anymore.

Computer scientists contributed significantly to the firm establishment of the information-processing dogma by using expressions such as "intelligence," "memory," and "language" to describe computers, which led most people—including the scientists themselves—to think that these terms refer to the well-known human phenomena. This, however, is a grave misunderstanding, which has helped to perpetuate, and even reinforce, the Cartesian image of human beings as machines.

Recent developments in cognitive science have made it clear that human intelligence is utterly different from machine, or "artificial," intelligence. The human nervous system does not process any information (in the sense of discrete elements existing ready-made in the outside world, to be picked up by the cognitive system), but interacts with the environment by continually modulating its structure.[31] Moreover, neuroscientists have discovered strong evidence that human intelligence, human memory, and human decisions are never completely rational but are always colored by emotions, as we all know from experience.[32] Our thinking is always accompanied by bodily sensations and processes. Even if we often tend to suppress these, we always think *also* with our body; and since computers do not have such a body, truly human problems will always be foreign to their intelligence.

These considerations imply that certain tasks should never be left to computers, as Joseph Weizenbaum asserted emphatically in his classic book, *Computer Power and Human Reason.* These tasks include all those that require genuine human qualities such as wisdom, compassion, respect, understanding, or love. Decisions and communications that require those qualities will dehumanize our lives if they are made by computers. To quote Weizenbaum:

> A line dividing human and machine intelligence must be drawn. If there is no such line, then advocates of computerized psychotherapy may be merely the heralds of an age in which man has finally been recognized as nothing but clockwork. . . . The very asking of the question, "What does a judge (or psychiatrist) know that we cannot tell a computer?" is a monstrous obscenity.[33]

Impact on Society

Because of its link with mechanistic science and its strong connections to the military, cybernetics enjoyed a very high prestige among the scientific establishment right from the beginning. Over the years this prestige increased further as computers spread rapidly throughout all strata of industrial society, bringing about profound changes in every area of our lives. Norbert Wiener predicted those changes, which have often been compared to a second industrial revolution, during the early years of cybernetics. More than that, he clearly perceived the shadow side of the new technologies he had helped to create:

> Those of us who have contributed to the new science of cybernetics . . . stand in a moral position which is, to say the least, not very comfortable. We have contributed to the initiation of a new science which . . . embraces technical developments with great possibilities for good and for evil.[34]

> Let us remember that the automatic machine . . . is the precise economic equivalent of slave labor. Any labor which competes with slave labor must accept the economic conditions of slave labor. It is perfectly clear that this will produce an unemployment situation in comparison with which the present recession and even the depression of the thirties will seem a pleasant joke.[35]

It is evident from these and other similar passages in Wiener's writings that he showed much more wisdom and foresight in his assessment of the social impact of computers than his successors. Today, forty years later, computers and the many other "information technologies" developed in the meantime are rapidly becoming autonomous and totalitarian, redefining our basic concepts and eliminating alternative worldviews. As Neil Postman, Jerry Mander, and other technology critics have shown, this is typical of the "megatechnologies" that have come to dominate industrial societies around the world.[36] Increasingly, all forms of culture are being subordinated to technology, and technological innovation, rather than the increase in human well-being, has become synonymous with progress.

The spiritual impoverishment and loss of cultural diversity through excessive use of computers is especially serious in the field of education. As Neil Postman put it succinctly, "When a computer is used for learning, the meaning of 'learning' is changed."[37] The use of computers in education is often praised as a revolution that will transform virtually every facet of the educational process. This view is promoted vigorously by the powerful computer industry, which encourages teachers to use computers as educational tools at all levels—even in kindergarten and preschool!—without ever mentioning the many harmful effects that may result from these irresponsible practices.[38]

The use of computers in schools is based on the now outdated view of human beings as information processors, which continually reinforces erroneous mechanistic concepts of thinking, knowledge, and communication. Information is presented as the basis of thinking, whereas in reality the human mind thinks with ideas, not with information. As Theodore Roszak shows in detail in *The Cult of Information,* information does not create ideas; ideas create information. Ideas are integrating patterns that derive not from information but from experience.[39]

In the computer model of cognition, knowledge is seen as context and value free, based on abstract data. But all meaningful knowledge is contextual knowledge, and much of it is tacit and experiential. Similarly, language is seen as a conduit through which "objective" information is communicated. In reality, as C. A. Bowers has argued eloquently, language is metaphoric, conveying tacit understandings shared within a culture.[40] In this connection it is also important to note that the language used by computer scientists and engineers is full of metaphors derived from the military—"command," "escape," "fail-safe," "pilot," "target," and so on—which introduce cultural biases, reinforce stereotypes, and inhibit certain groups, including most young, school-age girls, from fully participating in the learning experience.[41] A related issue of concern is the connection between computers and the violence and militaristic nature of most computer-based video games.

After dominating brain research and cognitive science for thirty years and creating a paradigm for technology that is still widespread today, the information-processing dogma was finally questioned seriously.[42] Critical arguments had been presented already during the pioneering phase of cybernetics. For example, it was argued that in actual brains there are no rules; there is no central logical processor, and information is not stored locally. Brains seem to operate on the basis of massive connectivity, storing information distributively and manifesting a self-organizing capacity that is nowhere to be found in computers. However, these alternative ideas were eclipsed in favor of the dominant computational view, until they reemerged thirty years later during the 1970s, when systems thinkers became fascinated by a new phenomenon with an evocative name—self-organization.

PART THREE

The Pieces of the Puzzle

5

Models of Self-Organization

Applied Systems Thinking

During the 1950s and 1960s systems thinking had a strong influence on engineering and management, where systems concepts—including those of cybernetics—were applied to solve practical problems. These applications gave rise to the new disciplines of systems engineering, systems analysis, and systemic management.[1]

As industrial enterprises became increasingly complex with the development of new chemical, electronic, and communications technologies, managers and engineers had to be concerned not only with large numbers of individual components, but also with the effects arising from the mutual interactions of those components, both in physical and organizational systems. Thus many engineers and project managers in large companies began to formulate strategies and methodologies that explicitly used systems concepts. Passages such as the following were found in many of the books on systems engineering that were published during the 1960s:

> The systems engineer must also be capable of predicting the emergent properties of the system, those properties, that is, which are possessed by the system but not its parts.[2]

The method of strategic thinking known as "systems analysis" was pioneered by the RAND Corporation, a military research and development institution founded in the late 1940s, which became the model for numerous "think tanks" specializing in policy making and the brokerage of technology.[3] Systems analysis grew out of operations research, the analysis and planning of military operations during World War II. These included the coordination of radar use with antiaircraft operations, the very same problems that also initiated the theoretical developments of cybernetics.

During the 1950s systems analysis went beyond military applications and became a broad systemic approach to cost-benefit analysis, involving mathematical models to examine a range of alternative programs designed to meet a well-defined goal. In the words of a popular text, published in 1968:

> One strives to look at the entire problem, as a whole, in context, and to compare alternative choices in the light of their possible outcomes.[4]

Soon after the development of systems analysis as a method for tackling complex organizational problems in the military, managers began to use the new approach to solve similar problems in business. "Systems-oriented management" became a new catchword, and during the 1960s and 1970s a whole series of books on management were published that featured the word "systems" in their titles.[5] The modeling technique of "systems dynamics," developed by Jay Forrester, and the "management cybernetics" of Stafford Beer are examples of comprehensive early formulations of the systems approach to management.[6]

A decade later a similar but much more subtle approach to management was developed by Hans Ulrich at the St. Gallen Business School in Switzerland.[7] Ulrich's approach is widely known in European management circles as the "St. Gallen model." It is based on the view of the business organization as a living social system and over the years has incorporated many ideas from biology, cognitive science, ecology, and evolutionary theory. These more recent developments gave rise to the new discipline of "systemic management," which is now taught at European business schools and advocated by management consultants.[8]

The Rise of Molecular Biology

While the systems approach had a significant influence on management and engineering during the 1950s and 1960s, its influence on biology, paradoxically, was almost negligible during that time. The 1950s were the decade of the spectacular triumph of genetics, the elucidation of the physical structure of DNA, which has been hailed as the greatest discovery in biology since Darwin's theory of evolution. For several decades this triumphal success totally eclipsed the systems view of life. Once again the pendulum swung back to mechanism.

The achievements of genetics brought about a significant shift in biological research, a new perspective that still dominates our academic institutions today. Whereas cells were regarded as the basic building blocks of living organisms during the nineteenth century, the attention shifted from cells to molecules toward the middle of the twentieth century, when geneticists began to explore the molecular structure of the gene.

Advancing to ever smaller levels in their explorations of the phenomena of life, biologists found that the characteristics of all living organisms—from bacteria to humans—were encoded in their chromosomes in the same chemical substance, using the same code script. After two decades of intensive research, the precise details of this code were unraveled. Biologists had discovered the alphabet of a truly universal language of life.[9]

This triumph of molecular biology resulted in the widespread belief that all biological functions can be explained in terms of molecular structures and mechanisms. Thus most biologists have become fervent reductionists, concerned with molecular details. Molecular biology, originally a small branch of the life sciences, has now become a pervasive and exclusive way of thinking that has led to a severe distortion of biological research.

At the same time, the problems that resist the mechanistic approach of molecular biology became ever more apparent during the second half of the century. While biologists know the precise structure of a few genes, they know very little of the ways in which genes communicate and cooperate in the development of an organism. In other words, they know the alphabet of the genetic code but have almost no idea of its syntax. It is now apparent that most of the DNA—perhaps as much as 95 percent—may be used for integrative activities about which biologists are likely to remain ignorant as long as they adhere to mechanistic models.

Critique of Systems Thinking

By the mid-1970s the limitations of the molecular approach to the understanding of life were evident. However, biologists saw little else on the horizon. The eclipse of systems thinking from pure science had become so complete that it was not considered a viable alternative. In fact, systems theory began to be seen as an intellectual failure in several critical essays. Robert Lilienfeld, for example, concluded his excellent account, *The Rise of Systems Theory,* published in 1978, with the following devastating critique:

> Systems thinkers exhibit a fascination for definitions, conceptualizations, and programmatic statements of a vaguely benevolent, vaguely moralizing nature. . . . They collect analogies between the phenomena of one field and those of another . . . the description of which seems to offer them an esthetic delight that is its own justification. . . . No evidence that systems theory has been used to achieve the solution of any substantive problem in any field whatsoever has appeared.[10]

The last part of this critique is definitely no longer justified today, as we shall see in the subsequent chapters of this book, and it may have been too harsh even in the 1970s. It could be argued even then that the understanding of living organisms as energetically open but organizationally closed systems, the recognition of feedback as the essential mechanism of homeostasis, and the cybernetic models of neural processes—to name just three examples that were well established at the time—represented major advances in the scientific understanding of life.

However, Lilienfeld was right in the sense that no formal systems theory of the kind envisaged by Bogdanov and Bertalanffy had been applied successfully in any field. Bertalanffy's goal, to develop his general systems theory into "a mathematical discipline, in itself purely formal but applicable to the various empirical sciences," was certainly never achieved.

The main reason for this "failure" was the lack of mathematical techniques for dealing with the complexity of living systems. Bogdanov and Bertalanffy both recognized that in open systems the simultaneous interactions of many variables generate the patterns of organization characteristic of life, but they lacked the means to describe the emergence of those patterns mathematically. Technically speaking, the mathematics of their time was limited to linear equations, which are inappropriate to describe the highly nonlinear nature of living systems.[11]

The cyberneticists concentrated on nonlinear phenomena like feedback loops and neural networks, and they had the beginnings of a corresponding nonlinear mathematics, but the real breakthrough came several decades later and was linked closely to the development of a new generation of powerful computers.

While the systemic approaches developed during the first half of the century did not result in a formal mathematical theory, they created a certain way of thinking, a new language, new concepts, and a whole intellectual climate that has led to significant scientific advances in recent years. Instead of a formal *systems theory* the decade of the 1980s saw the development of a series of successful *systemic models* that describe various aspects of the phenomenon of life. From these models the outlines of a coherent theory of living systems, together with the proper mathematical language, are now finally emerging.

The Importance of Pattern

The recent advances in our understanding of living systems are based on two developments that originated in the late 1970s, during the same years when Lilienfeld and others were writing their critiques of systems thinking. One was the discovery of the new mathematics of complexity, which is discussed in the following chapter. The other was the emergence of a powerful novel concept, that of self-organization, which had been implicit in the early discussions of the cyberneticists but was not developed explicitly for another thirty years.

To understand the phenomenon of self-organization, we first need to understand the importance of pattern. The idea of a pattern of organization—a configuration of relationships characteristic of a particular system—became the explicit focus of systems thinking in cybernetics and has been a crucial concept ever since. From the systems point of view, the understanding of life begins with the understanding of pattern.

We have seen that throughout the history of Western science and philosophy there has been a tension between the study of substance and the study of form.[12] The study of substance starts with the question, What is it made of?; the study of form with the question, What is its pattern? These are two very different approaches, which have been in competition with one another throughout our scientific and philosophical tradition.

The study of substance began in Greek antiquity in the sixth century B.C., when Thales, Parmenides, and other philosophers asked: What is reality made of? What are the ultimate constituents of matter? What is its essence? The answers to these questions define the various schools of the early era of Greek philosophy. Among them was the idea of four fundamental elements—earth, air, fire, water. In modern times those were recast into the chemical elements, now more than 100 but still a finite number of ultimate elements out of which all matter was thought to be made. Then Dalton identified the elements with atoms, and with the rise of atomic and nuclear physics in the twentieth century the atoms were further reduced to subatomic particles.

Similarly, in biology the basic elements were first organisms, or species, and in the eighteenth century biologists developed elaborate classification schemes for plants and animals. Then, with the discovery of cells as the common elements in all organisms, the focus shifted from organisms to cells. Finally, the cell was broken down into its macromolecules—enzymes, proteins, amino acids, and so forth—and molecular biology became the new frontier of research. In all those endeavors the basic question had not changed since Greek antiquity: What is reality made of? What are its ultimate constituents?

At the same time, throughout the same history of philosophy and science the study of pattern was always present. It began with the Pythagoreans in Greece and was continued by the alchemists, the Romantic poets, and various other intellectual movements. However, for most of the time the study of pattern was eclipsed by the study of substance until it reemerged forcefully in our century, when it was recognized by systems thinkers as essential to the understanding of life.

I shall argue that the key to a comprehensive theory of living systems lies in the synthesis of those two very different approaches, the study of substance (or structure) and the study of form (or pattern). In the study of structure we measure and weigh things. Patterns, however, cannot be measured or weighed; they must be mapped. To understand a pattern we must map a configuration of relationships. In other words, structure involves quantities, while pattern involves qualities.

The study of pattern is crucial to the understanding of living systems because systemic properties, as we have seen, arise from a configuration of ordered relationships.[13] Systemic properties are properties of a pattern. What is destroyed when a living organism is dissected is its pattern. The components are still there, but the configuration of relationships among them—the pattern—is destroyed, and thus the organism dies.

Most reductionist scientists cannot appreciate critiques of reductionism, because they fail to grasp the importance of pattern. They affirm that all living organisms are ultimately made of the same atoms and molecules that are the components of inorganic matter and that the laws of biology can therefore be reduced to those of physics and chemistry. While it is true that all living organisms are ultimately made of atoms and molecules, they are not "nothing but" atoms and molecules. There is something else to life, something nonmaterial and irreducible—a pattern of organization.

Networks—the Patterns of Life

Having appreciated the importance of pattern for the understanding of life, we can now ask: Is there a common pattern of organization that can be identified in all living systems? We shall see that this is indeed the case. This pattern of organization, common to all living systems, will be discussed in detail below.[14] Its most important property is that it is a network pattern. Whenever we encounter living systems—organisms, parts of organisms, or communities of organisms—we can observe that their components are arranged in network fashion. Whenever we look at life, we look at networks.

This recognition came into science in the 1920s, when ecologists began to study food webs. Soon after that, recognizing the network as the general pattern of life, systems thinkers extended network models to all systems levels. Cyberneticists, in particular, tried to understand the brain as a neural network and developed special mathematical techniques to analyze its patterns. The structure of the human brain is enormously complex. It contains about 10 billion nerve cells (neurons), which are interlinked in a vast network through 1,000 billion junctions (synapses). The whole brain can be divided into subsections, or subnetworks, which communicate with each other in network fashion. All this results in intricate patterns of intertwined webs, networks nesting within larger networks.[15]

The first and most obvious property of any network is its nonlinearity—it goes in all directions. Thus the relationships in a network pattern are nonlinear relationships. In particular, an influence, or message, may travel along a cyclical path, which may become a feedback loop. The concept of feedback is intimately connected with the network pattern.[16]

Because networks of communication may generate feedback loops, they may acquire the ability to regulate themselves. For example, a community that maintains an active network of communication will learn from its mistakes, because the consequences of a mistake will spread through the network and return to the source along feedback loops. Thus the community can correct its mistakes, regulate itself, and organize itself. Indeed, self-organization has emerged as perhaps *the* central concept in the systems view of life, and like the concepts of feedback and self-regulation, it is linked closely to networks. The pattern of life, we might say, is a network pattern capable of self-organization. This is a simple definition, yet it is based on recent discoveries at the very forefront of science.

Emergence of Self-Organization Concept

The concept of self-organization originated in the early years of cybernetics, when scientists began to construct mathematical models representing the logic inherent in neural networks. In 1943 the neuroscientist Warren McCulloch and the mathematician Walter Pitts published a pioneering paper entitled "A Logical Calculus of the Ideas Immanent in Nervous Activity," in which they showed that the logic of any physiological process, of any behavior, can be transformed into rules for constructing a network.[17]

In their paper the authors introduced idealized neurons represented by binary switching elements—in other words, elements that can switch "on" or "off"—and they modeled the nervous system as complex networks of those binary switching elements. In such a McCulloch-Pitts network the "on-off" nodes are coupled to one another in such a way that the activity of each node is governed by the prior activity of other nodes according to some "switching rule." For example, a node may switch on at the next moment only if a certain number of adjacent nodes are "on" at this moment. McCulloch and Pitts were able to show that although binary networks of this kind are simplified models, they are a good approximation of the networks embedded in the nervous system.

In the 1950s scientists began to actually build models of such binary networks, including some with little lamps flickering on and off at the nodes. To their great amazement they discovered that after a short time of random flickering, some ordered patterns would emerge in most networks. They would see waves of flickering pass through the network, or they would observe repeated cycles. Even though the initial state of the network was chosen at random, after a while those ordered patterns would emerge spontaneously, and it was that spontaneous emergence of order that became known as "self-organization."

As soon as this evocative term appeared in the literature, systems thinkers began to use it widely in different contexts. Ross Ashby in his early work was probably the first to describe the nervous system as "self-organizing."[18] The physicist and cybernetician Heinz von Foerster became a major catalyst for the self-organization idea in the late 1950s, organizing conferences around this topic, providing financial support for many of the participants, and publishing their contributions.[19]

For two decades Foerster maintained an interdisciplinary research group dedicated to the study of self-organizing systems. Centered at the Biological Computer Laboratory of the University of Illinois, this group was a close circle of friends and colleagues who worked away from the reductionist mainstream and whose ideas, being ahead of their time, were not widely published. However, those ideas were the seeds of many of the successful models of self-organizing systems developed during the late seventies and the eighties.

Heinz von Foerster's own contribution to the theoretical understanding of self-organization came very early and had to do with the concept of order. He asked: Is there a measure of order one could use to define the increase of order implied by "organization"? To solve this problem Foerster used the concept of "redundancy," defined mathematically in information theory by Claude Shannon, which measures the relative order of the system against the background of maximum disorder.[20]

Since then this approach has been superseded by the new mathematics of complexity, but in the late 1950s it allowed Foerster to develop an early qualitative model of self-organization in living systems. He coined the phrase "order from noise" to indicate that a self-organizing system does not just "import" order from its environment, but takes in energy-rich matter, integrates it into its own structure, and thereby increases its internal order.

During the seventies and eighties the key ideas of this early model were refined and elaborated by researchers in several countries who explored the phenomenon of self-organization in many different systems from the very small to the very large—Ilya Prigogine in Belgium, Hermann Haken and Manfred Eigen in Germany, James Lovelock in England, Lynn Margulis in the United States, Humberto Maturana and Francisco Varela in Chile.[21] The resulting models of self-organizing systems share certain key characteristics, which are the main ingredients of the emerging unified theory of living systems to be discussed in this book.

The first important difference between the early concept of self-organization in cybernetics and the more elaborate later models is that the latter include the creation of new structures and new modes of behavior in the self-organizing process. For Ashby all possible structural changes take place within a given "variety pool" of structures, and the survival chances of the system depend on the richness, or "requisite variety," of that pool. There is no creativity, no development, no evolution. The later models, by contrast, include the creation of novel structures and modes of behavior in the processes of development, learning, and evolution.

A second common characteristic of these models of self-organization is that they all deal with open systems operating far from equilibrium. A constant flow of energy and matter through the system is necessary for self-organization to take place. The striking emergence of new structures and new forms of behavior, which is the hallmark of self-organization, occurs only when the system is far from equilibrium.

The third characteristic of self-organization, common to all models, is the nonlinear interconnectedness of the system's components. Physically this nonlinear pattern results in feedback loops; mathematically it is described in terms of nonlinear equations.

Summarizing those three characteristics of self-organizing systems, we can say that self-organization is the spontaneous emergence of new structures and new forms of behavior in open systems far from equilibrium, characterized by internal feedback loops and described mathematically by nonlinear equations.

Dissipative Structures

The first, and perhaps most influential, detailed description of self-organizing systems was the theory of "dissipative structures" by the Russian-born chemist and physicist Ilya Prigogine, Nobel Laureate and professor of physical chemistry at the Free University of Brussels. Prigogine developed his theory from studies of physical and chemical systems, but according to his own recollections, he was led to do so after pondering the nature of life:

> I was very much interested in the problem of life. . . . I thought always that the existence of life is telling us something very important about nature.[22]

What intrigued Prigogine most was that living organisms are able to maintain their life processes under conditions of nonequilibrium. He became fascinated by systems far from thermal equilibrium and began an intensive investigation to find out under exactly what conditions nonequilibrium situations may be stable.

The crucial breakthrough occurred for Prigogine during the early 1960s, when he realized that systems far from equilibrium must be described by nonlinear equations. The clear recognition of this link between "far from equilibrium" and "nonlinearity" opened an avenue of research for Prigogine that would culminate a decade later in his theory of self-organization.

In order to solve the puzzle of stability far from equilibrium, Prigogine did not study living systems but turned to the much simpler phenomenon of heat convection, known as the "Bénard instability," which is now regarded as a classical case of self-organization. At the beginning of the century the French physicist Henri Bénard discovered that the heating of a thin layer of liquid may result in strangely ordered structures. When the liquid is uniformly heated from below, a constant heat flux is established, moving from the bottom to the top. The liquid itself remains at rest, and the heat is transferred by conduction alone. However, when the temperature difference between the top and bottom surfaces reaches a certain critical value, the heat flux is replaced by heat convection, in which the heat is transferred by the coherent motion of large numbers of molecules.

At this point a very striking ordered pattern of hexagonal

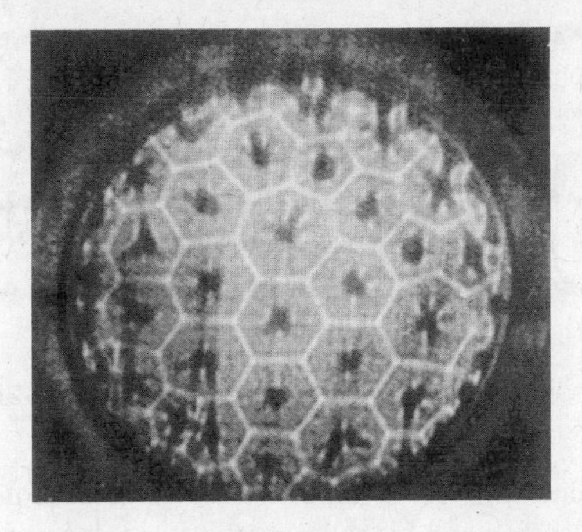

Figure 5-1
Pattern of hexagonal Bénard cells in a cylindrical container, viewed from above. The diameter of the container is approximately 10cm, the depth of the liquid approximately 0.5cm; from Bergé (1981).

("honeycomb") cells appears, in which hot liquid rises through the center of the cells, while the cooler liquid descends to the bottom along the cell walls (see figure 5-1). Prigogine's detailed analysis of these "Bénard cells" showed that as the system moves farther away from equilibrium (that is, from a state with uniform temperature throughout the liquid), it reaches a critical point of instability, at which the ordered hexagonal pattern emerges.[23]

The Bénard instability is a spectacular example of spontaneous self-organization. The nonequilibrium that is maintained by the continual flow of heat through the system generates a complex spatial pattern in which millions of molecules move coherently to form the hexagonal convection cells. Bénard cells, moreover, are not limited to laboratory experiments but also occur in nature in a wide variety of circumstances. For example, the flow of warm air from the surface of the earth toward outer space may generate hexagonal circulation vortices that leave their imprints on sand dunes in the desert and on arctic snow fields.[24]

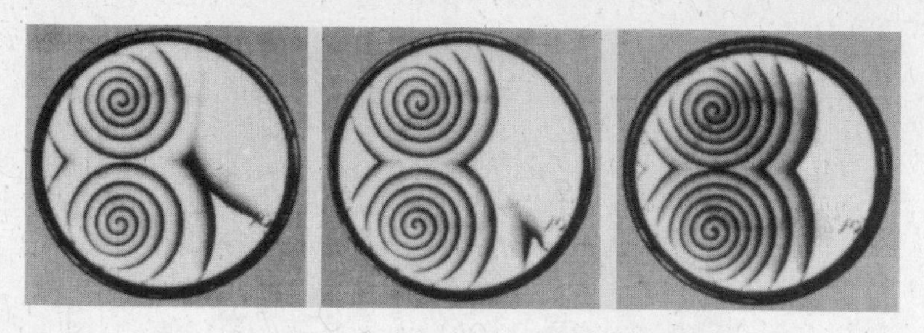

Figure 5-2
Wavelike chemical activity in the so-called Belousov-Zhabotinskii reaction; from Prigogine (1980).

Another amazing self-organization phenomenon studied extensively by Prigogine and his colleagues in Brussels are the so-called chemical clocks. These are reactions far from chemical equilibrium, which produce very striking periodic oscillations.[25] For example, if there are two kinds of molecules in the reaction, one "red" and one "blue," the system will be all blue at a certain point; then change its color abruptly to red; then again to blue; and so on at regular intervals. Different experimental conditions may also produce waves of chemical activity (see figure 5-2).

To change color all at once, the chemical system has to act as a whole, producing a high degree of order through the coherent activity of billions of molecules. Prigogine and his colleagues discovered that, as in the Bénard convection, this coherent behavior emerges spontaneously at critical points of instability far from equilibrium.

During the 1960s Prigogine developed a new nonlinear thermodynamics to describe the self-organization phenomenon in open systems far from equilibrium. "Classical thermodynamics," he explains, "leads to the concept of 'equilibrium structures' such as crystals. Bénard cells are structures too, but of a quite different nature. That is why we have introduced the notion of 'dissipative structures,' to emphasize the close association, at first paradoxical, in such situations between structure and order on the one side, and dissipation . . . on the other."[26] In classical thermodynamics the dissipation of energy in heat transfer, friction, and the like was always associated with waste. Prigogine's concept of a dissipative structure introduced a radical change in this view by showing that in open systems dissipation becomes a source of order.

In 1967 Prigogine presented his concept of dissipative structures for the first time in a lecture at a Nobel Symposium in Stockholm,[27] and four years later he published the first formulation of the full theory together with his colleague Paul Glansdorff.[28] According to Prigogine's theory, dissipative structures not only maintain themselves in a stable state far from equilibrium, but may even evolve. When the flow of energy and matter through them increases, they may go through new instabilities and transform themselves into new structures of increased complexity.

Prigogine's detailed analysis of this striking phenomenon showed that while dissipative structures receive their energy from outside, the instabilities and jumps to new forms of organization are the result of fluctuations amplified by positive feedback loops. Thus amplifying "runaway" feedback, which had always been regarded as destructive in cybernetics, appears as a source of new order and complexity in the theory of dissipative structures.

Laser Theory

During the early sixties, at the time when Ilya Prigogine realized the crucial importance of nonlinearity for the description of self-organizing systems, the physicist Hermann Haken in Germany had a very similar realization while studying the physics of lasers, which had just been invented. In a laser, certain special conditions combine to produce a transition from normal lamplight, which consists of an "incoherent" (unordered) mixture of light waves of different frequencies and phases, to "coherent" laser light consisting of one single, continuous, monochromatic wave train.

The high coherence of laser light is brought about by the coordination of light emissions from the individual atoms in the laser.

Haken recognized that this coordinated emission, resulting in the spontaneous emergence of coherence, or order, is a process of self-organization and that a nonlinear theory is needed to describe it properly. "In those days I had a lot of arguments with several American theorists," Haken remembers, "who were also working on lasers, but with a linear theory, and who did not realize that something qualitatively new is happening at this point."[29]

When the laser phenomenon was discovered, it was interpreted as an amplification process, which Einstein had already described in the early days of quantum theory. Atoms emit light when they are "excited"—that is, when their electrons have been lifted to higher orbits. After a while the electrons will spontaneously jump back to lower orbits and in the process emit energy in the form of wavelets of light. A beam of ordinary light consists of an incoherent mixture of these tiny wavelets emitted by individual atoms.

Under special circumstances, however, a passing light wave can "stimulate"—or, as Einstein called it, "induce"—an excited atom to emit its energy in such a way that the light wave is amplified. This amplified wave can, in turn, stimulate another atom to amplify it further, and eventually there will be an avalanche of amplifications. The resulting phenomenon was called "light amplification through stimulated emission of radiation," which gave rise to the acronym LASER.

The problem with this description is that different atoms in the laser material will simultaneously generate different light avalanches that are incoherent relative to each other. How then, Haken asked, do these unordered waves combine to produce a single coherent wave train? He was led to the answer by observing that a laser is a many-particle system far from thermal equilibrium.[30] It needs to be "pumped" from the outside to excite the atoms, which then radiate energy. Thus there is a constant flow of energy through the system.

While studying this phenomenon intensely during the 1960s, Haken found several parallels to other systems far from equilibrium, which led him to speculate that the transition from normal light to laser light might be an example of the self-organization processes that are typical of systems far from equilibrium.[31]

Haken coined the term "synergetics" to indicate the need for a new field of systematic study of those processes, in which the combined actions of many individual parts, such as the laser atoms, produce a coherent behavior of the whole. In an interview given in 1985 Haken explained:

> In physics, there is the term "cooperative effects," but it is used mainly for systems in thermal equilibrium. . . . I felt I should coin a term for cooperation [in] systems far from thermal equilibrium. . . . I wanted to emphasize that we need a new discipline for those processes. . . . So, one could see synergetics as a science dealing, perhaps not exclusively, with the phenomenon of self-organization.[32]

In 1970 Haken published his full nonlinear laser theory in the prestigious German physics encyclopedia *Handbuch der Physik*.[33] Treating the laser as a self-organizing system far from equilibrium, he showed that the laser action sets in when the strength of the external pumping reaches a certain critical value. Due to a special arrangement of mirrors on both ends of the laser cavity, only light emitted very close to the direction of the laser axis can remain in the cavity long enough to bring about the amplification process, while all other wave trains are eliminated.

Haken's theory makes it clear that although the laser needs to be pumped energetically from the outside to remain in a state far from equilibrium, the coordination of emissions is carried out by the laser light itself; it is a process of self-organization. Thus Haken arrived independently at a precise description of a self-organizing phenomenon of the kind Prigogine would call a dissipative structure.

The predictions of laser theory have been verified in great detail, and due to the pioneering work of Hermann Haken, the laser has become an important tool for the study of self-organization. At a symposium honoring Haken's sixtieth birthday, his collaborator Robert Graham paid an eloquent tribute to his work:

> It is one of Haken's great contributions to recognize that lasers are not only extremely important technological tools, but also highly

> interesting physical systems in themselves, which can teach us important lessons. . . . Lasers occupy a very interesting place between the quantum world and the classical world, and Haken's theory tells us how these worlds can be connected. . . . The laser can be seen at the crossroads between quantum and classical physics, between equilibrium and non-equilibrium phenomena, between phase transitions and self-organization, and between regular and chaotic dynamics. At the same time, it is a system which we understand both on a microscopic quantum mechanical and a macroscopic classical level. It is a solid ground for discovering general concepts of non-equilibrium physics.[34]

Hypercycles

Whereas Prigogine and Haken were led to the concept of self-organization by studying physical and chemical systems that go through points of instability and generate new forms of order, the biochemist Manfred Eigen used the same concept to shed light on the puzzle of the origin of life. According to standard Darwinian theory, living organisms formed out of "molecular chaos" by chance through random mutations and natural selection. However, it has often been pointed out that the probability of even simple cells to emerge in this way during the known age of the Earth is vanishingly small.

Manfred Eigen, Nobel Laureate in chemistry and director of the Max Planck Institute for Physical Chemistry in Göttingen, proposed in the early seventies that the origin of life on Earth may have been the result of a process of progressive organization in chemical systems far from equilibrium, involving "hypercycles" of multiple feedback loops. Eigen, in effect, postulated a prebiological phase of evolution, in which selection processes occur in the molecular realm "as a material property inherent in special reaction systems,"[35] and he coined the term "molecular self-organization" to describe these prebiological evolutionary processes.[36]

The special reaction systems studied by Eigen are known as "catalytic cycles." A catalyst is a substance that increases the rate of a chemical reaction without itself being changed in the process.

Catalytic reactions are crucial processes in the chemistry of life. The most common and most efficient catalysts are the enzymes, which are essential components of cells promoting vital metabolic processes.

When Eigen and his colleagues studied catalytic reactions involving enzymes in the 1960s, they observed that in biochemical systems far from equilibrium, i.e., systems exposed to energy flows, different catalytic reactions combine to form complex networks that may contain closed loops. Figure 5-3 shows an example of such a catalytic network, in which fifteen enzymes catalyze each other's formations in such a way that a closed loop, or catalytic cycle, is formed.

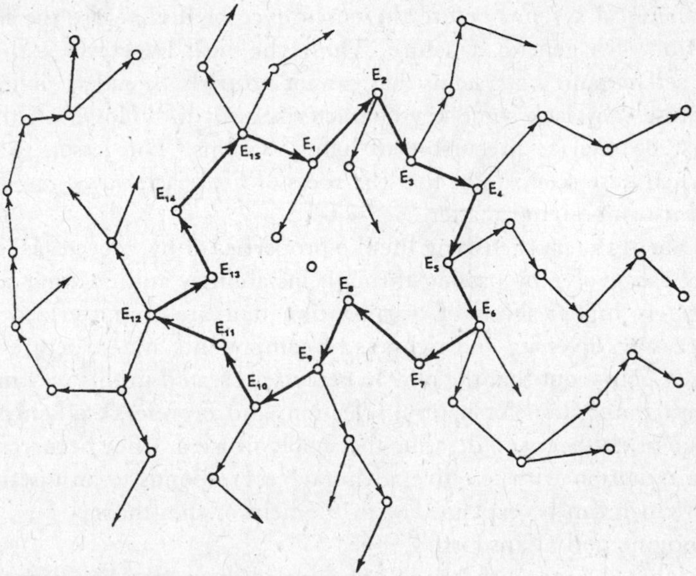

Figure 5-3
A catalytic network of enzymes, including a closed loop (E1 . . . E15); from Eigen (1971).

These catalytic cycles are at the core of self-organizing chemical systems such as the chemical clocks studied by Prigogine, and they also play an essential role in the metabolic functions of living organisms. They are remarkably stable and can persist under a wide range of conditions.[37] Eigen discovered that with sufficient time and a continuing flow of energy, catalytic cycles tend to interlock to form closed loops in which the enzymes produced in one cycle act as catalysts in the subsequent cycle. He coined the term "hypercycles" for those loops in which each link is a catalytic cycle.

Hypercycles turn out to be not only remarkable stable, but also capable of self-replication and of correcting replication errors, which means that they can conserve and transmit complex information. Eigen's theory shows that such self-replication—which is, of course, well-known for living organisms—may have occurred in chemical systems before the emergence of life, before the formation of a genetic structure. These chemical hypercycles, then, are self-organizing systems that cannot properly be called "living" because they lack some key characteristics of life. However, they must be seen as precursors to living systems. The lesson to be learned here seems to be that the roots of life reach down into the realm of nonliving matter.

One of the most striking lifelike properties of hypercycles is that they can evolve by passing through instabilities and creating successively higher levels of organization that are characterized by increasing diversity and richness of components and structures.[38] Eigen points out that the new hypercycles created in this way may be in competition for natural selection, and he refers explicitly to Prigogine's theory to describe the whole process: "The occurrence of a mutation with selective advantage corresponds to an instability, which can be explained with the help of the [theory] . . . of Prigogine and Glansdorff."[39]

Manfred Eigen's theory of hypercycles shares the key concepts of self-organization with Ilya Prigogine's theory of dissipative structures and Hermann Haken's laser theory—the state of the system far from equilibrium; the development of amplification processes through positive feedback loops; and the appearance of instabilities leading to the creation of new forms of organization. In addition, Eigen made the revolutionary step of using a Darwinian approach to describe evolutionary phenomena at a prebiological, molecular level.

Autopoiesis—the Organization of the Living

The hypercycles studied by Eigen self-organize, self-reproduce, and evolve. Yet one hesitates to call these cycles of chemical reactions "alive." What properties, then, must a system have to be called truly living? Can we make a clear distinction between living and nonliving systems? What is the precise connection between self-organization and life?

These were the questions the Chilean neuroscientist Humberto Maturana asked himself during the 1960s. After six years of studies and research in biology in England and the United States, where he collaborated with Warren McCulloch's group at MIT and was strongly influenced by cybernetics, Maturana returned to the University of Santiago in 1960. There he specialized in neuroscience and, in particular, in the understanding of color perception.

From this research two major questions crystallized in Maturana's mind. As he remembered it later, "I entered a situation in which my academic life was divided, and I oriented myself in search of the answers to two questions that seemed to lead in opposite directions, namely: 'What is the organization of the living?' and 'What takes place in the phenomenon of perception?' "[40]

Maturana struggled with these questions for almost a decade, and it was his genius to find a common answer to both of them. In so doing, he made it possible to unify two traditions of systems thinking that had been concerned with phenomena on different sides of the Cartesian division. While organismic biologists had explored the nature of biological form, cyberneticists had attempted to understand the nature of mind. Maturana realized in the late sixties that the key to both of these puzzles lay in the understanding of "the organization of the living."

In the fall of 1968 Maturana was invited by Heinz von Foerster to join his interdisciplinary research group at the University of

Illinois and to participate in a symposium on cognition held in Chicago a few months later. This gave him an ideal opportunity to present his ideas on cognition as a biological phenomenon.[41]

What, then, was Maturana's central insight? In his own words:

> My investigations of color perception led me to a discovery that was extraordinarily important for me: The nervous system operates as a closed network of interactions, in which every change of the interactive relations between certain components always results in a change of the interactive relations of the same or of other components."[42]

From this discovery Maturana drew two conclusions, which gave him the answers to his two major questions. He hypothesized that the "circular organization" of the nervous system is the basic organization of all living systems: "Living systems . . . [are] organized in a closed causal circular process that allows for evolutionary change in the way the circularity is maintained, but not for the loss of the circularity itself."[43]

Since all changes in the system take place within this basic circularity, Maturana argued that the components that specify the circular organization must also be produced and maintained by it. And he concluded that this network pattern, in which the function of each component is to help produce and transform other components while maintaining the overall circularity of the network, is the basic "organization of the living."

The second conclusion Maturana drew from the circular closure of the nervous system amounted to a radically new understanding of cognition. He postulated that the nervous system is not only self-organizing but also continually self-referring, so that perception cannot be viewed as the representation of an external reality but must be understood as the continual creation of new relationships within the neural network: "The activities of nerve cells do not reflect an environment independent of the living organism and hence do not allow for the construction of an absolutely existing external world."[44]

According to Maturana, perception and, more generally, cognition do not *represent* an external reality, but rather *specify* one through the nervous system's process of circular organization. From this premise Maturana then took the radical step of postulating that the process of circular organization itself—with or without a nervous system—is identical to the process of cognition:

> Living systems are cognitive systems, and living as a process is a process of cognition. This statement is valid for all organisms, with and without a nervous system.[45]

This way of identifying cognition with the process of life itself is indeed a radically new conception. Its implications are far-reaching and will be discussed in detail in the following pages.[46]

After publishing his ideas in 1970, Maturana began a long collaboration with Francisco Varela, a younger neuroscientist at the University of Santiago who was Maturana's student before he became his collaborator. According to Maturana, their collaboration began when Varela challenged him in a conversation to find a more formal and more complete description for the concept of circular organization.[47] They immediately set to work on a complete verbal description of Maturana's idea before attempting to construct a mathematical model, and they began by inventing a new name for it—*autopoiesis*.

Auto, of course, means "self" and refers to the autonomy of self-organizing systems; and *poiesis*—which shares the same Greek root as the word "poetry"—means "making." So *autopoiesis* means "self-making." Since they had coined a new word without a history, it was easy to use it as a technical term for the distinctive organization of living systems. Two years later Maturana and Varela published their first description of autopoiesis in a long essay,[48] and by 1974 they and their colleague Ricardo Uribe had developed a corresponding mathematical model for the simplest autopoietic system, the living cell.[49]

Maturana and Varela begin their essay on autopoiesis by characterizing their approach as "mechanistic" to distinguish it from vitalist approaches to the nature of life: "Our approach will be mechanistic: no forces or principles will be adduced which are not found in the physical universe." However, the next sentence makes it immediately clear that the authors are not Cartesian mechanists but systems thinkers:

> Yet, our problem is the living organization and therefore our interest will not be in properties of components, but in processes and relations between processes realized through components.[50]

They go on to refine their position with the important distinction between "organization" and "structure," which had been an implicit theme during the entire history of systems thinking but was not addressed explicitly until the development of cybernetics.[51] Maturana and Varela make the distinction crystal clear. The organization of a living system, they explain, is the set of relations among its components that characterize the system as belonging to a particular class (such as a bacterium, a sunflower, a cat, or a human brain). The description of that organization is an abstract description of relationships and does not identify the components. The authors assume that autopoiesis is a general pattern of organization, common to all living systems, whichever the nature of their components.

The structure of a living system, by contrast, is constituted by the actual relations among the physical components. In other words, the system's structure is the physical embodiment of its organization. Maturana and Varela emphasize that the system's organization is independent of the properties of its components, so that a given organization can be embodied in many different manners by many different kinds of components.

Having clarified that their concern is with organization, not structure, the authors then proceed to define autopoiesis, the organization common to all living systems. It is a network of production processes, in which the function of each component is to participate in the production or transformation of other components in the network. In this way the entire network continually "makes itself." It is produced by its components and in turn produces those components. "In a living system," the authors explain, "the product of its operation is its own organization."[52]

An important characteristic of living systems is that their autopoietic organization includes the creation of a boundary that specifies the domain of the network's operations and defines the system as a unit. The authors point out that catalytic cycles, in particular, do not constitute living systems, because their boundary is determined by factors (such as a physical container) that are independent of the catalytic processes.

It is also interesting to note that physicist Geoffrey Chew formulated his so-called bootstrap hypothesis about the composition and interactions of subatomic particles, which sounds quite similar to the concept of autopoiesis, about a decade before Maturana first published his ideas.[53] According to Chew, strongly interacting particles, or "hadrons," form a network of interactions in which "each particle helps to generate other particles, which in turn generate it."[54]

However, there are two key differences between the hadron bootstrap and autopoiesis. Hadrons are potential "bound states" of each other in the probabilistic sense of quantum theory, which does not apply to Maturana's "organization of the living." Moreover, a network of subatomic particles interacting through high-energy collisions cannot be said to be autopoietic because it does not form any boundary.

According to Maturana and Varela, the concept of autopoiesis is necessary and sufficient to characterize the organization of living systems. However, this characterization does not include any information about the physical constitution of the system's components. To understand the properties of the components and their physical interactions, a description of the system's structure in the language of physics and chemistry must be added to the abstract description of its organization. The clear distinction between these two descriptions—one in terms of structure and the other in terms of organization—makes it possible to integrate structure-oriented models of self-organization (such as those by Prigogine and Haken) and organization-oriented models (as those by Eigen and Maturana-Varela) into a coherent theory of living systems.[55]

Gaia—the Living Earth

The key ideas underlying the various models of self-organizing systems just described crystallized within a few years during the early 1960s. In the United States Heinz von Foerster assembled his interdisciplinary research group and held several conferences on self-organization; in Belgium Ilya Prigogine realized the crucial link between nonequilibrium systems and nonlinearity; in Germany Hermann Haken developed his nonlinear laser theory and Manfred Eigen worked on catalytic cycles; and in Chile Humberto Maturana puzzled over the organization of living systems.

At the same time, the atmospheric chemist James Lovelock had an illuminating insight that led him to formulate a model that is perhaps the most surprising and most beautiful expression of self-organization—the idea that the planet Earth as a whole is a living, self-organizing system.

The origins of Lovelock's daring hypothesis lie in the early days of the NASA space program. While the idea of the Earth being alive is very ancient and speculative theories about the planet as a living system had been formulated several times,[56] the space flights during the early 1960s enabled human beings for the first time to actually look at our planet from outer space and perceive it as an integrated whole. This perception of the Earth in all its beauty—a blue-and-white globe floating in the deep darkness of space—moved the astronauts deeply and, as several have since declared, was a profound spiritual experience that forever changed their relationship to the Earth.[57] The magnificent photographs of the whole Earth that they brought back provided the most powerful symbol for the global ecology movement.

While the astronauts looked at the planet and beheld its beauty, the environment of the Earth was also examined from outer space by the sensors of scientific instruments, and so were the environments of the moon and the nearby planets. During the 1960s the Soviet and American space programs launched over fifty space probes, most of them to explore the moon but some traveling beyond to Venus and Mars.

At that time NASA invited James Lovelock to the Jet Propulsion Laboratories in Pasadena, California, to help them design instruments for the detection of life on Mars.[58] NASA's plan was to send a spacecraft to Mars that would search for life at the landing site by performing a series of experiments with the Martian soil. While Lovelock worked on technical problems of instrument design, he also asked himself a more general question: How can we be sure that the Martian way of life, if any, will reveal itself to tests based on Earth's lifestyle? Over the following months and years this question led him to think deeply about the nature of life and how it could be recognized.

In contemplating this problem, Lovelock found that the fact that all living organisms take in energy and matter and discard waste products was the most general characteristic of life he could identify. Much like Prigogine, he thought that one should be able to express this key characteristic mathematically in terms of entropy, but then his reasoning went in a different direction. Lovelock assumed that life on any planet would use the atmosphere and oceans as fluid media for raw materials and waste products. Therefore, he speculated, one might be able, somehow, to detect the existence of life by analyzing the chemical composition of a planet's atmosphere. Thus if there was life on Mars, the Martian atmosphere should reveal some special combination of gases, some characteristic "signature" that could be detected even from Earth.

These speculations were confirmed dramatically when Lovelock and a colleague, Dian Hitchcock, began a systematic analysis of the Martian atmosphere, using observations made from Earth, and compared it with a similar analysis of the Earth's atmosphere. They discovered that the chemical compositions of the two atmospheres are strikingly different. While there is very little oxygen, a lot of carbon dioxide (CO_2), and no methane in the Martian atmosphere, the Earth's atmosphere contains massive amounts of oxygen, almost no CO_2, and a lot of methane.

Lovelock realized that the reason for that particular atmospheric profile on Mars is that on a planet with no life, all possible chemical reactions among the gases in the atmosphere were completed a long time ago. Today no more chemical reactions are possible on Mars; there is complete chemical equilibrium in the Martian atmosphere.

The situation on Earth is exactly the opposite. The terrestrial atmosphere contains gases like oxygen and methane, which are very likely to react with each other but coexist in high proportions, resulting in a mixture of gases far from chemical equilibrium. Lovelock realized that this special state must be due to the presence of life on Earth. Plants produce oxygen constantly and other organisms produce other gases, so that the atmospheric gases are being replenished continually while they undergo chemical reactions. In other words, Lovelock recognized the Earth's atmosphere as an open system, far from equilibrium, characterized by a constant flow of energy and matter. His chemical analysis identified the very hallmark of life.

This insight was so momentous for Lovelock that he still remembers the exact moment when it occurred:

> For me, the personal revelation of Gaia came quite suddenly—like a flash of enlightenment. I was in a small room on the top floor of a building at the Jet Propulsion Laboratory in Pasadena, California. It was the autumn of 1965 . . . and I was talking with a colleague, Dian Hitchcock, about a paper we were preparing. . . . It was at that moment that I glimpsed Gaia. An awesome thought came to me. The Earth's atmosphere was an extraordinary and unstable mixture of gases, yet I knew that it was constant in composition over quite long periods of time. Could it be that life on Earth not only made the atmosphere, but also regulated it—keeping it at a constant composition, and at a level favorable for organisms?[59]

The process of self-regulation is the key to Lovelock's idea. He knew from astrophysics that the heat of the sun has increased by 25 percent since life began on Earth and that, in spite of this increase, the Earth's surface temperature has remained constant, at a level comfortable for life, during those four billion years. What if the Earth were able to regulate its temperature, he asked, as well as other planetary conditions—the composition of its atmosphere, the salinity of its oceans, and so on—just as living organisms are able to self-regulate and keep their body temperature and other variables constant? Lovelock realized that this hypothesis amounted to a radical break with conventional science:

> Consider Gaia theory as an alternative to the conventional wisdom that sees the Earth as a dead planet made of inanimate rocks, ocean, and atmosphere, and merely inhabited by life. Consider it as a real system, comprising all of life and all of its environment tightly coupled so as to form a self-regulating entity.[60]

The space scientists at NASA, by the way, did not like Lovelock's discovery at all. They had developed an impressive array of life-detection experiments for their Viking mission to Mars, and now Lovelock was telling them that there was really no need to send a spacecraft to the red planet in search of life. All they needed was a spectral analysis of the Martian atmosphere, which could easily be done through a telescope on Earth. Not surprisingly, NASA disregarded Lovelock's advice and continued to develop the Viking program. Their spacecraft landed on Mars several years later, and as Lovelock had predicted, it found no trace of life.

In 1969, at a scientific meeting in Princeton, Lovelock for the first time presented his hypothesis of the Earth as a self-regulating system.[61] Shortly after that a novelist friend, recognizing that Lovelock's idea represents the renaissance of a powerful ancient myth, suggested the name "Gaia hypothesis" in honor of the Greek goddess of the Earth. Lovelock gladly accepted the suggestion and in 1972 published the first extensive version of his idea in a paper titled "Gaia as Seen through the Atmosphere."[62]

At that time Lovelock had no idea *how* the Earth might regulate its temperature and the composition of its atmosphere, except that he knew that the self-regulating processes had to involve organisms in the biosphere. Nor did he know which organisms produced which gases. At the same time, however, the American microbiologist Lynn Margulis was studying the very processes Lovelock needed to understand—the production and removal of gases by various organisms, including especially the myriad bacteria in the Earth's soil. Margulis remembers that she kept asking,

"Why does everybody agree that atmospheric oxygen . . . comes from life, but no one speaks about the other atmospheric gases coming from life?"[63] Soon several of her colleagues recommended that she speak to James Lovelock, which led to a long and fruitful collaboration that resulted in the full scientific Gaia hypothesis.

The scientific backgrounds and areas of expertise of James Lovelock and Lynn Margulis turned out to be a perfect match. Margulis had no problem answering Lovelock's many questions about the biological origins of atmospheric gases, while Lovelock contributed concepts from chemistry, thermodynamics, and cybernetics to the emerging Gaia theory. Thus the two scientists were able gradually to identify a complex network of feedback loops that—so they hypothesized—bring about the self-regulation of the planetary system.

The outstanding feature of these feedback loops is that they link together living and nonliving systems. We can no longer think of rocks, animals, and plants as being separate. Gaia theory shows that there is a tight interlocking between the planet's living parts—plants, microorganisms, and animals—and its nonliving parts—rocks, oceans, and the atmosphere.

The carbon dioxide cycle is a good illustration of this point.[64] The Earth's volcanoes have spewed out huge amounts of carbon dioxide (CO_2) for millions of years. Since CO_2 is one of the main greenhouse gases, Gaia needs to pump it out of the atmosphere; otherwise it would get too hot for life. Plants and animals recycle massive amounts of CO_2 and oxygen in the processes of photosynthesis, respiration, and decay. However, these exchanges are always in balance and do not affect the level of CO_2 in the atmosphere. According to Gaia theory, the excess of carbon dioxide in the atmosphere is removed and recycled by a vast feedback loop, which involves rock weathering as a key ingredient.

In the process of rock weathering, rocks combine with rainwater and carbon dioxide to form various chemicals, called carbonates. The CO_2 is thus taken out of the atmosphere and bound in liquid solutions. These are purely chemical processes that do not require the participation of life. However, Lovelock and others discovered that the presence of soil bacteria vastly increases the

Figure 5-4
Oceanic alga (coccolithophore) with chalk shell.

rate of rock weathering. In a sense, these soil bacteria act as catalysts for the process of rock weathering, and the entire carbon dioxide cycle could be viewed as the biological equivalent of the catalytic cycles studied by Manfred Eigen.

The carbonates are then washed down into the ocean, where tiny algae, invisible to the naked eye, absorb them and use them to make exquisite shells of chalk (calcium carbonate). So the CO_2 that was in the atmosphere has now ended up in the shells of those minute algae (figure 5-4). In addition, ocean algae also absorb carbon dioxide directly from the air.

When the algae die, their shells rain down to the ocean floor, where they form massive sediments of limestone (another form of calcium carbonate). Because of their enormous weight, the limestone sediments gradually sink into the mantle of the Earth and melt and may even trigger the movements of tectonic plates. Eventually some of the CO_2 contained in the molten rocks is spewed out again by volcanoes and sent on another round in the great Gaian cycle.

The entire cycle—linking volcanoes to rock weathering, to soil bacteria, to oceanic algae, to limestone sediments, and back to volcanoes—acts as a giant feedback loop, which contributes to the regulation of the Earth's temperature. As the sun gets hotter, bacterial action in the soil is stimulated, which increases the rate of rock weathering. This in turn pumps more CO_2 out of the atmosphere and thus cools the planet. According to Lovelock and Margulis, similar feedback cycles—interlinking plants and rocks, animals and atmospheric gases, microorganisms and the oceans—regulate the Earth's climate, the salinity of its oceans, and other important planetary conditions.

Gaia theory looks at life in a systemic way, bringing together geology, microbiology, atmospheric chemistry, and other disciplines whose practitioners are not used to communicating with each other. Lovelock and Margulis challenged the conventional view that those are separate disciplines, that the forces of geology set the conditions for life on Earth and that the plants and animals were mere passengers who by chance found just the right conditions for their evolution. According to Gaia theory, life creates the conditions for its own existence. In the words of Lynn Margulis:

> Simply stated, the [Gaia] hypothesis says that the surface of the Earth, which we've always considered to be the *environment* of life, is really *part* of life. The blanket of air—the troposphere—should be considered a circulatory system, produced and sustained by life. . . . When scientists tell us that life adapts to an essentially passive environment of chemistry, physics, and rocks, they perpetuate a severely distorted view. Life actually makes and forms and changes the environment to which it adapts. Then that "environment" feeds back on the life that is changing and acting and growing in it. There are constant cyclical interactions.[65]

At first the resistance of the scientific community to this new view of life was so strong that the authors found it impossible to publish their hypothesis. Established academic journals, such as *Science* and *Nature,* turned it down. Finally the astronomer Carl Sagan, who served as editor of the journal *Icarus,* invited Lovelock and Margulis to publish the Gaia hypothesis in his journal.[66] It is intriguing that of all the theories and models of self-organization, the Gaia hypothesis encountered by far the strongest resistance. One is tempted to wonder whether this highly irrational reaction by the scientific establishment was triggered by the evocation of Gaia, the powerful archetypal myth.

Indeed, the image of Gaia as a sentient being was the main implicit argument for the rejection of the Gaia hypothesis after its publication. Scientists expressed it by claiming that the hypothesis was unscientific because it was teleological—that is, implying the idea of natural processes being shaped by a purpose. "Neither Lynn Margulis nor I have ever proposed that planetary self-regulation is purposeful," Lovelock protests. "Yet we have met persistent, almost dogmatic, criticism that our hypothesis is teleological."[67]

This criticism harks back to the old debate between mechanists and vitalists. While mechanists hold that all biological phenomena will eventually be explained in terms of the laws of physics and chemistry, vitalists postulate the existence of a nonphysical entity, a causal agent directing the life processes that defy mechanistic explanations.[68] Teleology—from the Greek *telos* ("purpose")—asserts that the causal agent postulated by vitalism is purposeful, that there is purpose and design in nature. By strenuously opposing vitalist and teleological arguments, the mechanists still struggle with the Newtonian metaphor of God as a clockmaker. The currently emerging theory of living systems has finally overcome the debate between mechanism and teleology. As we shall see, it views living nature as mindful and intelligent without the need to assume any overall design or purpose.[69]

The representatives of mechanistic biology attacked the Gaia hypothesis as teleological, because they could not imagine how life on Earth could create and regulate the conditions for its own existence without being conscious and purposeful. "Are there committee meetings of species to negotiate next year's temperature?" those critics asked with malicious humor.[70]

Lovelock responded with an ingenious mathematical model, called "Daisyworld." It represents a vastly simplified Gaian system, in which it is absolutely clear that the temperature regulation is an emergent property of the system that arises automatically, without any purposeful action, as a consequence of feedback loops between the planet's organisms and their environment.[71]

Daisyworld is a computer model of a planet, warmed by a sun with steadily increasing heat radiation and with only two species growing on it—black daisies and white daisies. Seeds of these daisies are scattered throughout the planet, which is moist and fertile everywhere, but daisies will grow only within a certain temperature range.

Lovelock programmed his computer with the mathematical equations corresponding to all these conditions, chose a planetary temperature at the freezing point for the starting condition, and then let the model run on the computer. "Will the evolution of the Daisyworld ecosystem lead to the self-regulation of climate?" was the crucial question he asked himself.

The results were spectacular. As the model planet warms up, at some point the equator becomes warm enough for plant life. The black daisies appear first because they absorb heat better than the white daisies and are therefore more fit for survival and reproduction. Thus in its first phase of evolution Daisyworld shows a ring of black daisies scattered around the equator (figure 5-5).

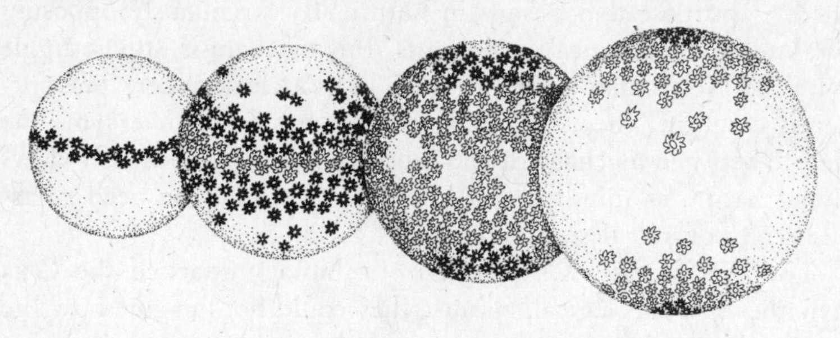

Figure 5-5
The four evolutionary phases of Daisyworld.

As the planet warms up further, the equator becomes too hot for the black daisies to survive and they begin to colonize the subtropical zones. At the same time, white daisies appear around the equator. Because they are white, they reflect heat and cool themselves, which allows them to survive better in hot zones than the black daisies. In the second phase, then, there is a ring of white daisies around the equator and the subtropical and temperate zones are filled with black daisies, while it is still too cold around the poles for any daisies to grow.

Then the sun gets hotter still and plant life becomes extinct at the equator, where it is now too hot even for the white daisies. In the meantime white daisies have replaced the black daisies in the temperate zones, and black daisies are beginning to appear around the poles. Thus the third phase shows the equator bare, the temperate zones populated with white daisies, and the zones around the poles filled with black daisies with just the pole caps themselves without any plant life. In the last phase, finally, vast regions around the equator and the subtropical zones are too hot for any daisies to survive, while there are white daisies in the temperate zones and black daisies at the poles. After that it becomes too hot on the model planet for any daisies to grow and all life becomes extinct.

These are the basic dynamics of the Daisyworld system. The crucial property of the model that brings about self-regulation is that the black daisies, by absorbing heat, warm not only themselves but also the planet. Similarly, while the white daisies reflect heat and cool themselves, they also cool the planet. Thus heat is absorbed and reflected throughout the evolution of Daisyworld, depending on which species of daisies are present.

When Lovelock plotted the changes of temperature on the planet throughout its evolution, he got the striking result that the planetary temperature is kept constant throughout the four phases (figure 5-6). When the sun is relatively cold, Daisyworld increases its own temperature through heat absorption by the black daisies; as the sun gets hotter, the temperature is lowered gradually because of the progressive predominance of heat-reflecting white daisies. Thus Daisyworld, without any foresight or planning, "regulates its own temperature over a vast time range by the dance of the daisies."[72]

Feedback loops that link environmental influences to the growth of daisies, which in turn affect the environment, are an essential feature of the Daisyworld model. When this cycle is broken so that there is no influence of the daisies on the environment, the daisy populations fluctuate wildly and the whole system goes chaotic. But as soon as the loops are closed by linking the daisies

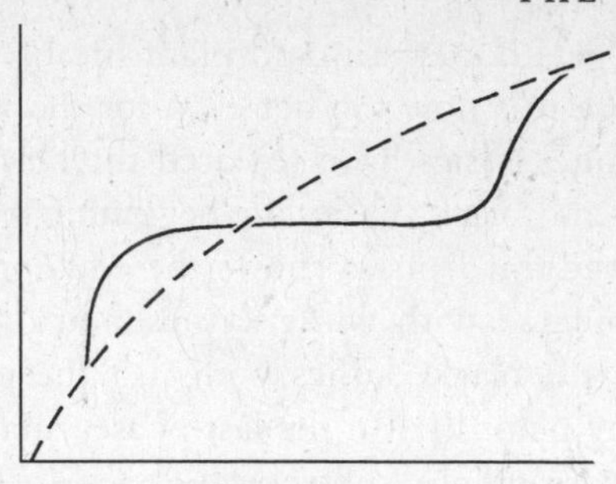

Figure 5-6
Evolution of temperature on Daisyworld: The dashed curve shows the rise of temperature with no life present; the solid curve shows how life maintains a constant temperature; from Lovelock (1991).

back to the environment, the model stabilizes and self-regulation occurs.

Since then Lovelock has designed much more sophisticated versions of Daisyworld. Instead of just two, there are many species of daisies with varying pigments in the new models; there are models in which the daisies evolve and change color; models in which rabbits eat the daisies and foxes eat the rabbits; and so on.[73] The net result of these highly complex models is that the small temperature fluctuations that were present in the original Daisyworld model have flattened out, and self-regulation becomes more and more stable as the model's complexity increases. In addition, Lovelock put catastrophes into his models, which wipe out 30 percent of the daisies at regular intervals. He found that Daisyworld's self-regulation is remarkably resilient under these severe disturbances.

All these models generated lively discussions among biologists, geophysicists, and geochemists, and since they were first published the Gaia hypothesis has gained much more respect in the scientific community. In fact, there are now several research teams in various parts of the world who work on detailed formulations of the Gaia theory.[74]

An Early Synthesis

In the late 1970s, almost twenty years after the key criteria of self-organization were discovered in various contexts, detailed mathematical theories and models of self-organizing systems had been formulated, and a set of common characteristics became apparent—the continual flow of energy and matter through the system, the stable state far from equilibrium, the emergence of new patterns of order, the central role of feedback loops, and the mathematical description in terms of nonlinear equations.

At that time the Austrian physicist Erich Jantsch, then at the University of California at Berkeley, presented an early synthesis of the new models of self-organization in a book titled *The Self-Organizing Universe,* which was based mainly on Prigogine's theory of dissipative structures.[75] Although Jantsch's book is now largely outdated, because it was written before the new mathematics of complexity became widely known and because it did not include the full concept of autopoiesis as the organization of living systems, it was of tremendous value at the time. It was the first book that made Prigogine's work available to a broad audience, and it attempted to integrate a large number of then very new concepts and ideas into a coherent paradigm of self-organization. My own synthesis of these concepts in the present book is, in a sense, a reformulation of Erich Jantsch's earlier work.

6

The Mathematics of Complexity

The view of living systems as self-organizing networks whose components are all interconnected and interdependent has been expressed repeatedly, in one way or another, throughout the history of philosophy and science. However, detailed models of self-organizing systems could be formulated only very recently when new mathematical tools became available that allowed scientists to model the nonlinear interconnectedness characteristic of networks. The discovery of this new "mathematics of complexity" is increasingly being recognized as one of the most important events in twentieth-century science.

The theories and models of self-organization described in the previous pages deal with highly complex systems involving thousands of interdependent chemical reactions. Over the past three decades a new set of concepts and techniques for dealing with that enormous complexity has emerged, one that is beginning to form a coherent mathematical framework. As yet there is no definitive name for this new mathematics. It is popularly known as "the mathematics of complexity" and technically as "dynamical systems theory," "systems dynamics," "complex dynamics," or "nonlinear dynamics." The term "dynamical systems theory" is perhaps the one most widely used.

To avoid confusion it is useful to keep in mind that dynamical systems theory is not a theory of physical phenomena but a mathematical theory whose concepts and techniques are applied to a broad range of phenomena. The same is true for chaos theory and the theory of fractals, which are important branches of dynamical systems theory.

The new mathematics, as we shall see in detail, is one of relationships and patterns. It is qualitative rather than quantitative and thus embodies the shift of emphasis that is characteristic of systems thinking—from objects to relationships, from quantity to quality, from substance to pattern. The development of large high-speed computers has played a crucial role in the new mastery of complexity. With their help mathematicians are now able to solve complex equations that had previously been intractable and to trace out the solutions as curves in a graph. In this way they have discovered new qualitative patterns of behavior of those complex systems, a new level of order underlying the seeming chaos.

Classical Science

To appreciate the novelty of the new mathematics of complexity it is instructive to contrast it with the mathematics of classical science. Science in the modern sense of the term began in the late sixteenth century with Galileo Galilei, who was the first to carry out systematic experiments and use mathematical language to formulate the laws of nature he discovered. At that time science was still called "natural philosophy," and when Galileo said "mathematics" he meant geometry. "Philosophy," he wrote, "is written in that great book which ever lies before our eyes; but we cannot understand it if we do not first learn the language and characters in which it is written. This language is mathematics, and the characters are triangles, circles, and other geometric figures."[1]

Galileo inherited this view from the philosophers of ancient Greece, who tended to geometrize all mathematical problems and to seek answers in terms of geometrical figures. Plato's Academy in Athens, the principal Greek school of science and philosophy for nine centuries, is said to have had a sign above its entrance, "Let no one enter here who is unacquainted with geometry."

Several centuries later a very different approach to solving mathematical problems, known as algebra, was developed by Islamic philosophers in Persia, who in turn had learned it from Indian mathematicians. The word is derived from the Arabic *al-jabr* ("binding together") and refers to the process of reducing the number of unknown quantities by binding them together in equations. Elementary algebra involves equations in which letters—by convention taken from the beginning of the alphabet—stand for various constant numbers. A well-known example, which most readers will remember from their school years, is this equation:

$$(a + b)^2 = a^2 + 2ab + b^2$$

Higher algebra involves relationships, called "functions," among unknown variable numbers, or "variables," which are denoted by letters taken by convention from the end of the alphabet. For example, in the equation

$$y = x + 1$$

the variable y is said to be "a function of x," which is written in mathematical shorthand as $y = f(x)$.

At the time of Galileo, then, there were two different approaches to solving mathematical problems, geometry and algebra, which came from different cultures. These two approaches were unified by René Descartes. A generation younger than Galileo, Descartes is usually regarded as the founder of modern philosophy, and he was also a brilliant mathematician. Descartes's invention of a method to make algebraic formulas and equations visible as geometric shapes was the greatest among his many contributions to mathematics.

The method, now known as analytic geometry, involves Cartesian coordinates, the coordinate system invented by Descartes and named after him. For example, when the relationship between the two variables x and y in our previous example, the equation $y = x + 1$, is pictured in a graph with Cartesian coordinates, we see

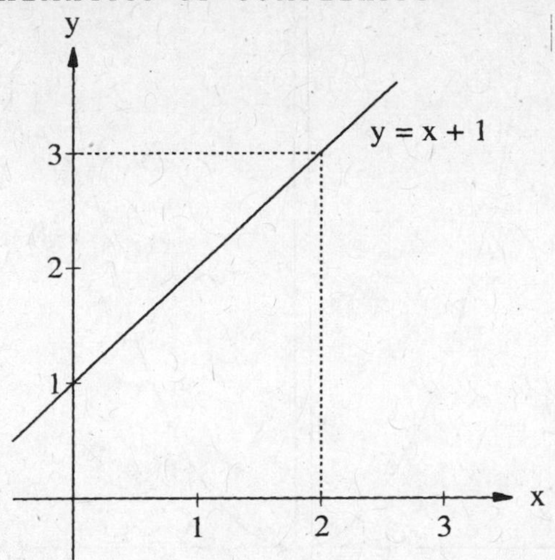

Figure 6-1
Graph corresponding to the equation $y = x + 1$. For any point on the straight line the value of the y-coordinate is always one unit more than that of the x-coordinate.

that it corresponds to a straight line (figure 6-1). This is why equations of this type are called "linear" equations.

Similarly, the equation $y = x^2$ is represented by a parabola (figure 6-2). Equations of this type, corresponding to curves in the Cartesian grid, are called "nonlinear" equations. They have the distinguishing feature that one or several of their variables are squared or raised to higher powers.

Differential Equations

With Descartes's new method, the laws of mechanics that Galileo had discovered could be expressed either in algebraic form as equations or in geometric form as visual shapes. However, there was a major mathematical problem, which neither Galileo nor Descartes nor any of their contemporaries could solve. They were unable to write down an equation describing the movement of a body at variable speed, accelerating or slowing down.

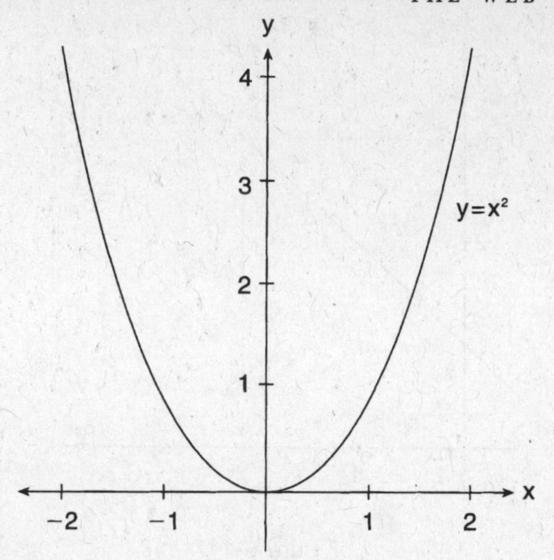

Figure 6-2
Graph corresponding to the equation $y = x^2$. For any point on the parabola the y-coordinate is equal to the square of the x-coordinate.

To understand the problem, let us consider two moving bodies, one traveling with constant speed, the other accelerating. If we plot their distance against time, we obtain the two graphs shown in figure 6-3. In the case of the accelerating body, the speed changes at every instant, and this is something Galileo and his contemporaries could not express mathematically. In other words, they were unable to calculate the exact speed of the accelerating body at a given time.

This was achieved a century later by Isaac Newton, the giant of classical science, and around the same time by the German philosopher and mathematician Gottfried Wilhelm Leibniz. To solve the problem that had plagued mathematicians and natural philosophers for centuries, Newton and Leibniz independently invented a new mathematical method, which is now known as calculus and is considered the gateway to "higher mathematics."

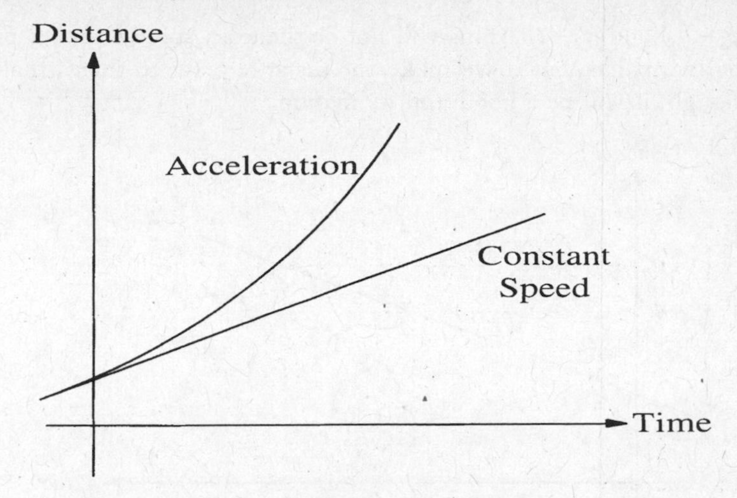

Figure 6-3
Graphs showing the motion of two bodies, one moving at constant speed, the other accelerating.

To see how Newton and Leibniz tackled the problem is very instructive and does not require any technical language. We all know how to calculate the speed of a moving body if it remains constant. If you drive 20 mph, this means that in one hour you will cover a distance of twenty miles, in two hours forty miles, and so on. Therefore, to obtain the speed of the car you simply divide the distance (e.g., forty miles) by the time it took you to cover that distance (e.g., two hours). In our graph this means that we have to divide the difference between two distance coordinates by the difference between two time coordinates, as shown in figure 6-4.

When the speed of the car varies, as it does in any real situation, of course, you will have driven more or less than twenty miles after one hour, depending on how often you accelerated and slowed down. How can we calculate the exact speed at a particular time in such a case?

Here is how Newton did it. He said, first let us calculate (in the example of accelerating motion) the approximate speed between two points by replacing the curve between them by a straight line. As shown in figure 6-5 the speed is again the ratio between

$(d_2 - d_1)$ and $(t_2 - t_1)$. This will not be the exact speed at either of the two points, but if we make the distance between them small enough, it will be a good approximation.

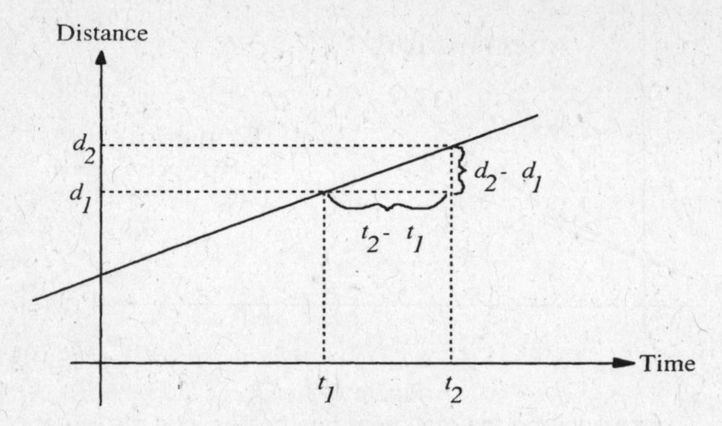

Figure 6-4
To calculate a constant speed, divide the difference between distance coordinates (d_2–d_1) by the difference between time coordinates (t_2–t_1).

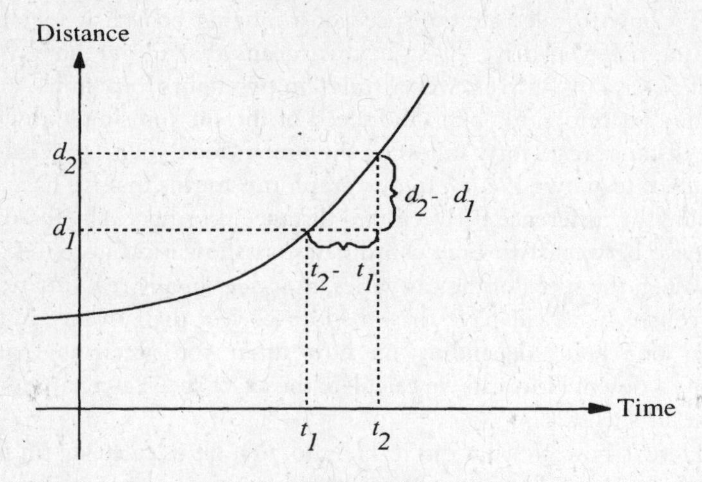

Figure 6-5
Calculating the approximate speed between two points in the case of accelerating motion.

And then Newton said, now let's shrink the triangle, which is formed by the curve and the coordinate differences, by moving the two points on the curve closer and closer together. As we do so. the straight line between the two points will come closer and closer to the curve, and the error in calculating the speed between the two points will be smaller and smaller. Finally, when we reach *the limit of infinitely small differences*—this is the crucial step!—the two points on the curve merge into one, and we get the exact speed at that point. Geometrically the straight line will then be a tangent to the curve.

To shrink this triangle to zero mathematically and calculate the ratio between two infinitely small differences is far from trivial. The precise definition of the limit of the infinitely small is the crux of the entire calculus. Technically an infinitely small difference is called a "differential," and the calculus invented by Newton and Leibniz is therefore known as differential calculus. Equations involving differentials are called differential equations.

For science, the invention of the differential calculus was a giant step. For the first time in human history the concept of the infinite, which had intrigued philosophers and poets from time immemorial, was given a precise mathematical definition, which opened countless new possibilities for the analysis of natural phenomena.

The power of this new analytical tool can be illustrated with the celebrated paradox of Zeno from the early Eleatic school of Greek philosophy. According to Zeno, the great athlete Achilles can never catch up with a tortoise in a race in which the tortoise is granted an initial lead. For when Achilles has completed the distance corresponding to that lead, the tortoise will have covered a farther distance; while Achilles covers that, the tortoise will have advanced again; and so on to infinity. Although the athlete's lag keeps decreasing, it will never disappear. At any given moment the tortoise will always be ahead. Therefore, Zeno concluded, Achilles, the fastest runner of antiquity, can never catch up with the tortoise.

Greek philosophers and their successors argued about this paradox for centuries, but they could never resolve it because the exact definition of the infinitely small eluded them. The flaw in Zeno's argument lies in the fact that even though it will take Achilles an infinite number of *steps* to reach the tortoise, this does not take an infinite *time.* With the tools of Newton's calculus it is easy to show that a moving body will run through an infinite number of infinitely small intervals in a finite time.

In the seventeenth century Isaac Newton used his calculus to describe all possible motions of solid bodies in terms of a set of differential equations, which have been known as "Newton's equations of motion" ever since. This feat was hailed by Einstein as "perhaps the greatest advance in thought that a single individual was ever privileged to make."[2]

Facing Complexity

During the eighteenth and nineteenth centuries the Newtonian equations of motion were cast into more general, more abstract, and more elegant forms by some of the greatest minds in the history of mathematics. Successive reformulations by Pierre Laplace, Leonhard Euler, Joseph Lagrange, and William Hamilton did not change the content of Newton's equations, but their increasing sophistication allowed scientists to analyze an ever-broadening range of natural phenomena.

Applying his theory to the movement of the planets, Newton himself was able to reproduce the basic features of the solar system, though not its finer details. Laplace, however, refined and perfected Newton's calculations to such an extent that he succeeded in explaining the motion of the planets, moons, and comets down to the smallest details, as well as the flow of the tides and other phenomena related to gravity.

Encouraged by this brilliant success of Newtonian mechanics in astronomy, physicists and mathematicians extended it to the motion of fluids and to the vibrations of strings, bells, and other elastic bodies, and again it worked. These impressive successes made scientists of the early nineteenth century believe that the universe was indeed a large mechanical system running according to the Newtonian laws of motion. Thus Newton's differential equations became the mathematical foundation of the mechanistic paradigm. The Newtonian world machine was seen as being completely causal and deterministic. All that happened had a definite cause and gave rise to a definite effect, and the future of any part of the system could—in principle—be predicted with absolute certainty if its state at any time was known in all details.

In practice, of course, the limitations of modeling nature through Newton's equations of motion soon became apparent. As the British mathematician Ian Stewart points out, "To *set up* the equations is one thing, to *solve* them quite another."[3] Exact solutions were restricted to a few simple and regular phenomena, while the complexity of vast areas of nature seemed to elude all mechanistic modeling. For example, the relative motion of two bodies under the force of gravity could be calculated precisely; that of three bodies was already too difficult for an exact solution; and when it came to gases with millions of particles, the situation seemed hopeless.

On the other hand, for a long time physicists and chemists had observed regularities in the behavior of gases, which had been formulated in terms of so-called gas laws—simple mathematical relations among the temperature, volume, and pressure of a gas. How could this apparent simplicity be derived from the enormous complexity of the motion of the individual molecules?

In the nineteenth century the great physicist James Clerk Maxwell found an answer. Even though the exact behavior of the molecules of a gas could not be determined, Maxwell argued that their *average* behavior might give rise to the observed regularities. Hence Maxwell proposed to use statistical methods to formulate the laws of motion for gases:

> The smallest portion of matter which we can subject to experiment consists of millions of molecules, none of which ever becomes individually sensible to us. We cannot, therefore, ascertain the actual motion of any of these molecules; so we are obliged to abandon the strict historical method, and to adopt the statistical method of dealing with large groups of molecules.[4]

Maxwell's method was indeed highly successful. It enabled physicists immediately to explain the basic properties of a gas in terms of the average behavior of its molecules. For example, it became clear that the pressure of a gas is the force caused by the molecules' average push,[5] while the temperature turned out to be proportional to their average energy of motion. Statistics and probability theory, its theoretical basis, had been developed since the seventeenth century and could readily be applied to the theory of gases. The combination of statistical methods with Newtonian mechanics resulted in a new branch of science, appropriately called "statistical mechanics," which became the theoretical foundation of thermodynamics, the theory of heat.

Nonlinearity

Thus, by the end of the nineteenth century scientists had developed two different mathematical tools to model natural phenomena—exact, deterministic equations of motion for simple systems; and the equations of thermodynamics, based on statistical analysis of average quantities, for complex systems.

Although these two techniques were quite different, they had one thing in common. They both featured *linear* equations. The Newtonian equations of motion are very general, appropriate for both linear and nonlinear phenomena; indeed, every now and then nonlinear equations were formulated. But since these were usually too complex to be solved, and because of the seemingly chaotic nature of the associated physical phenomena—such as turbulent flows of water and air—scientists generally avoided the study of nonlinear systems.[6]

So, whenever nonlinear equations appeared, they were immediately "linearized"—in other words, replaced by linear approximations. Thus instead of describing the phenomena in their full complexity, the equations of classical science deal with *small* oscillations, *shallow* waves, *small* changes of temperature, and so forth. As Ian Stewart observes, this habit became so ingrained that many equations were linearized *while they were being set up,* so that the science textbooks did not even include the full nonlinear versions. Consequently most scientists and engineers came to believe that virtually all natural phenomena could be described by linear equations. "As the world was a clockwork for the eighteenth century, it was a linear world for the 19th and most of the 20th century."[7]

The decisive change over the last three decades has been to recognize that nature, as Stewart puts it, is "relentlessly nonlinear." Nonlinear phenomena dominate much more of the inanimate world than we had thought, and they are an essential aspect of the network patterns of living systems. Dynamical systems theory is the first mathematics that enables scientists to deal with the full complexity of these nonlinear phenomena.

The exploration of nonlinear systems over the past decades has had a profound impact on science as a whole, as it has forced us to reevaluate some very basic notions about the relationships between a mathematical model and the phenomena it describes. One of those notions concerns our understanding of simplicity and complexity.

In the world of linear equations we thought we knew that systems described by simple equations behaved in simple ways, while those described by complicated equations behaved in complicated ways. In the nonlinear world—which includes most of the real world, as we begin to discover—simple deterministic equations may produce an unsuspected richness and variety of behavior. On the other hand, complex and seemingly chaotic behavior can give rise to ordered structures, to subtle and beautiful patterns. In fact, in chaos theory the term "chaos" has acquired a new technical meaning. The behavior of chaotic systems is not merely random but shows a deeper level of patterned order. As we shall see below, the new mathematical techniques enable us to make these underlying patterns visible in distinct shapes.

Another important property of nonlinear equations that has been disturbing to scientists is that exact prediction is often impossible, even though the equations may be strictly deterministic. We shall see that this striking feature of nonlinearity has brought about an important shift of emphasis from quantitative to qualitative analysis.

Feedback and Iterations

The third important property of nonlinear systems is a consequence of the frequent occurrence of self-reinforcing feedback processes. In linear systems small changes produce small effects, and large effects are due either to large changes or to a sum of many small changes. In nonlinear systems, by contrast, small changes may have dramatic effects because they may be amplified repeatedly by self-reinforcing feedback. Such nonlinear feedback processes are the basis of the instabilities and the sudden emergence of new forms of order that are so characteristic of self-organization.

Mathematically a feedback loop corresponds to a special kind of nonlinear process known as iteration (Latin for "repetition"), in which a function operates repeatedly on itself. For example, if the function consists in multiplying the variable x by 3—i.e., $f(x) = 3x$—the iteration consists in repeated multiplications. In mathematical shorthand this is written as follows:

$$x \rightarrow 3x$$
$$3x \rightarrow 9x$$
$$9x \rightarrow 27x$$
$$\text{etc.}$$

Each of these steps is called a "mapping." If we visualize the variable x as a line of numbers, the operation $x \rightarrow 3x$ maps each number to another number on the line. More generally, a mapping that consists in multiplying x by a constant number k is written like this:

$$x \rightarrow kx$$

An iteration found often in nonlinear systems, which is very simple and yet produces a wealth of complexity, is the mapping

$$x \rightarrow kx(1 - x)$$

where the variable x is restricted to values between 0 and 1. This mapping, known to mathematicians as "logistic mapping," has many important applications. It is used by ecologists to describe the growth of a population under opposing tendencies and is therefore also known as the "growth equation."[8]

Exploring the iterations of various logistic mappings is a fascinating exercise, which can easily be carried out with a small pocket calculator.[9] To see the essential feature of these iterations, let us choose again the value k = 3:

$$x \to 3x(1 - x)$$

The variable *x* can be visualized as a line segment running from 0 to 1, and it is easy to calculate the mappings for a few points, as follows:

$$\begin{aligned}
0 &\to 0(1 - 0) &&= 0 \\
0.2 &\to 0.6\,(1 - 0.2) &&= 0.48 \\
0.4 &\to 1.2\,(1 - 0.4) &&= 0.72 \\
0.6 &\to 1.8\,(1 - 0.6) &&= 0.72 \\
0.8 &\to 2.4\,(1 - 0.8) &&= 0.48 \\
1 &\to 3(1 - 1) &&= 0
\end{aligned}$$

When we mark these numbers on two line segments, we see that numbers between 0 and 0.5 are mapped to numbers between 0 and 0.75. Thus 0.2 becomes 0.48, and 0.4 becomes 0.72. Numbers between 0.5 and 1 are mapped to the same segment but in reverse order. Thus 0.6 becomes 0.72, and 0.8 becomes 0.48. The overall effect is shown in figure 6-6. We see that the mapping stretches the segment so that it covers the distance from 0 to 1.5 and then folds it back over itself, resulting in a segment running from 0 to 0.75 and back.

An iteration of this mapping will result in repeated stretching and folding operations, much like a baker stretches and folds a dough over and over again. The iteration is therefore called, very aptly, the "baker transformation." As the stretching and folding proceeds, neighboring points on the line segment will be moved farther and farther away from each other, and it is impossible to predict where a particular point will end up after many iterations.

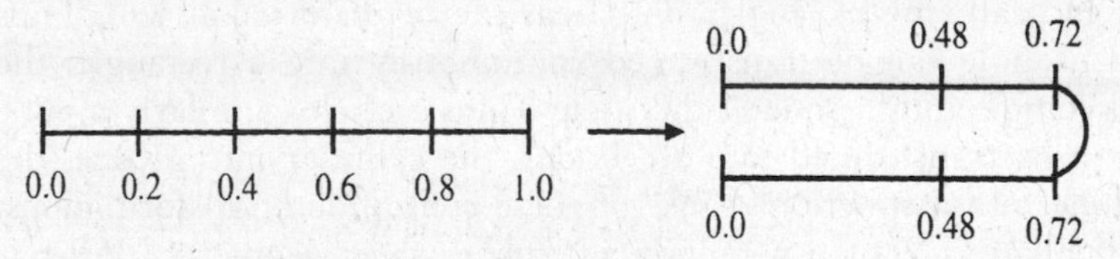

Figure 6-6
The logistic mapping, or "baker transformation."

Even the most powerful computers round off their calculations at a certain number of decimal points, and after a sufficient number of iterations even the most minute round-off errors will have added up to enough uncertainty to make predictions impossible. The baker transformation is a prototype of the nonlinear, highly complex, and unpredictable processes known technically as chaos.

Poincaré and the Footprints of Chaos

Dynamical systems theory, the mathematics that has made it possible to bring order into chaos, was developed very recently, but its foundations were laid at the turn of the century by one of the greatest mathematicians of the modern era, Jules Henri Poincaré. Among all the mathematicians of this century, Poincaré was the last great generalist. He made innumerable contributions in virtually all branches of mathematics. His collected works run into several hundred volumes.

From the vantage point of the late twentieth century we can see that Poincaré's greatest contribution was to bring visual imagery back into mathematics.[10] From the seventeenth century on, the style of European mathematics had shifted gradually from geometry, the mathematics of visual shapes, to algebra, the mathematics of formulas. Laplace, especially, was one of the great formalizers who boasted that his *Analytical Mechanics* contained no pictures. Poincaré reversed that trend, breaking the stranglehold of analysis and formulas that had become ever more opaque and turning once again to visual patterns.

Poincaré's visual mathematics, however, is not the geometry of Euclid. It is a geometry of a new kind, a mathematics of patterns and relationships known as topology. Topology is a geometry in which all lengths, angles, and areas can be distorted at will. Thus a triangle can be transformed continuously into a rectangle, the rectangle into a square, the square into a circle. Similarly a cube can be transformed into a cylinder, the cylinder into a cone, the cone into a sphere. Because of these continuous transformations, topology is known popularly as "rubber sheet geometry." All figures that can be transformed into each other by continuous bending, stretching, and twisting are called "topologically equivalent."

However, not everything is changeable by these topological transformations. In fact, topology is concerned precisely with those properties of geometric figures that do not change when the figures are transformed. Intersections of lines, for example, remain intersections, and the hole in a torus (doughnut) cannot be transformed away. Thus a doughnut may be transformed topologically into a coffee cup (the hole turning into a handle) but never into a pancake. Topology, then, is really a mathematics of relationships, of unchangeable, or "invariant," patterns.

Poincaré used topological concepts to analyze the qualitative features of complex dynamical problems and, in doing so, laid the foundations for the mathematics of complexity that would emerge a century later. Among the problems Poincaré analyzed in this way was the celebrated three-body problem in celestial mechanics—the relative motion of three bodies under their mutual gravitational attraction—which nobody had been able to solve.[11] By applying his topological method to a slightly simplified three-body problem, Poincaré was able to determine the general shape of its trajectories and found it to be of awesome complexity:

> When one tries to depict the figure formed by these two curves and their infinity of intersections . . . [one finds that] these intersections form a kind of net, web, or infinitely tight mesh; neither of the two curves can ever cross itself, but must fold back on itself in a very complex way in order to cross the links of the web infinitely many times. One is struck with the complexity of this figure that I am not even attempting to draw.[12]

What Poincaré pictured in his mind is now called a "strange attractor." In the words of Ian Stewart, "Poincaré was gazing at the footprints of chaos."[13]

By showing that simple deterministic equations of motion can produce unbelievable complexity that defies all attempts at prediction, Poincaré challenged the very foundations of Newtonian mechanics. However, because of a quirk of history, scientists at the turn of the century did not take up this challenge. A few years after Poincaré published his work on the three-body problem, Max Planck discovered energy quanta and Albert Einstein published his special theory of relativity.[14] For the next half century physicists and mathematicians were fascinated with the revolutionary developments in quantum physics and relativity theory, and Poincaré's groundbreaking discovery moved backstage. It was not until the 1960s that scientists stumbled again into the complexities of chaos.

Trajectories in Abstract Spaces

The mathematical techniques that have enabled researchers during the past three decades to discover ordered patterns in chaotic systems are based on Poincaré's topological approach and are closely linked to the development of computers. With the help of today's high-speed computers, scientists can solve nonlinear equations by techniques that were not available before. These powerful computers can easily trace out the complex trajectories that Poincaré did not even attempt to draw.

As most readers will remember from school, an equation is solved by manipulating it until you get a final formula as the solution. This is called solving the equation "analytically." The result is always a formula. Most nonlinear equations describing natural phenomena are too difficult to be solved analytically. But there is another way, which is called solving the equation "numerically." This involves trial and error. You try out various combinations of numbers for the variables until you find the ones that fit the equation. Special techniques and tricks have been developed for doing this efficiently, but for most equations the process is extremely cumbersome, takes a long time, and gives only very rough, approximate solutions.

All this changed when the new powerful computers arrived on the scene. Now we have programs for numerically solving an equation in extremely fast and accurate ways. With the new methods nonlinear equations can be solved to any degree of accuracy. However, the solutions are of a very different kind. The result is not a formula, but a large collection of values for the variables that satisfy the equation, and the computer can be programmed to trace out the solution as a curve, or set of curves, in a graph. This technique has enabled scientists to solve the complex nonlinear equations associated with chaotic phenomena and to discover order beneath the seeming chaos.

To reveal these ordered patterns, the variables of a complex system are displayed in an abstract mathematical space called "phase space." This is a well-known technique that was developed in thermodynamics at the turn of the century.[15] Every variable of the system is associated with a different coordinate in this abstract space. Let us illustrate this with a very simple example, a ball swinging back and forth on a pendulum. To describe the pendulum's motion completely, we need two variables: the angle, which can be positive or negative, and the velocity, which can again be positive or negative, depending on the direction of the swing. With these two variables, angle and velocity, we can describe the state of motion of the pendulum completely at any moment.

If we now draw a Cartesian coordinate system, in which one coordinate is the angle and the other the velocity (see figure 6-7), this coordinate system will span a two-dimensional space in which certain points correspond to the possible states of motion of the pendulum. Let us see where these points are. At the extreme elongations the velocity is zero. This gives us two points on the horizontal axis. At the center, where the angle is zero, the velocity is at its maximum, either positive (swinging one way) or negative

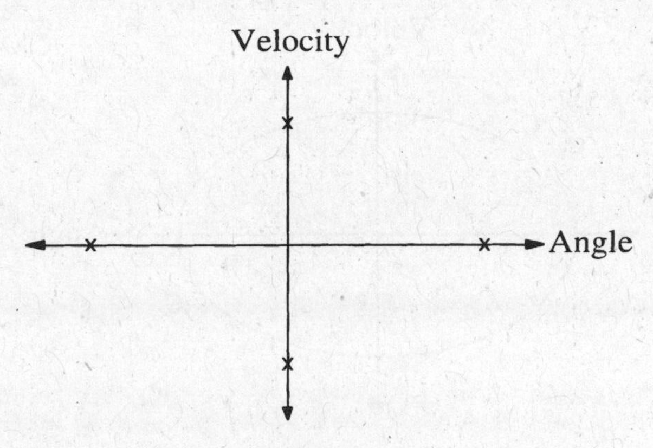

Figure 6-7
The two-dimensional phase space of a pendulum.

(swinging the other way). This gives us two points on the vertical axis. Those four points in phase space, which we have marked in figure 6-7, represent the extreme states of the pendulum—maximum elongation and maximum velocity. The exact location of these points will depend on our units of measurement.

If we were to go on and mark the points corresponding to the states of motion among the four extremes, we would find that they lie on a closed loop. We could make it a circle by choosing our units of measurement appropriately, but in general it will be some kind of an ellipse (figure 6-8). This loop is called the pendulum's trajectory in phase space. It completely describes the system's motion. All the variables of the system (two in our simple case) are represented by a single point, which will always be somewhere on this loop. As the pendulum swings back and forth, the point in phase space will go around the loop. At any moment we can measure the two coordinates of the point in phase space, and we will know the exact state—angle and velocity—of the system. Note that this loop is not in any sense a trajectory of the ball on the pendulum. It is a curve in an abstract mathematical space, composed of the system's two variables.

So this is the phase-space technique. The variables of the system are pictured in an abstract space, in which a single point describes

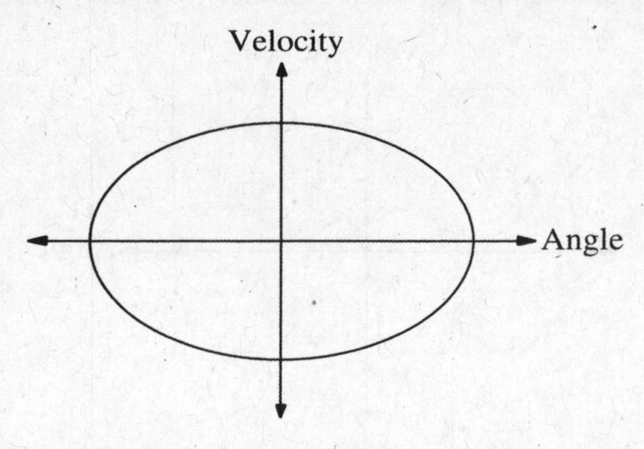

Figure 6-8
Trajectory of the pendulum in phase space.

the entire system. As the system changes, the point traces out a trajectory in phase space—a closed loop in our example. When the system is not a simple pendulum but much more complicated, it will have many more variables, but the technique is still the same. Each variable is represented by a coordinate in a different dimension in phase space. If there are sixteen variables, we will have a sixteen-dimensional space. A single point in that space will describe the state of the entire system completely, because this single point has sixteen coordinates, each corresponding to one of the system's sixteen variables.

Of course, we cannot visualize a phase space with sixteen dimensions; this is why it is called an abstract mathematical space. Mathematicians don't seem to have any problems with such abstractions. They are just as comfortable in spaces that cannot be visualized. At any rate, as the system changes, the point representing its state in phase space will move around in that space, tracing out a trajectory. Different initial states of the system correspond to different starting points in phase space and will, in general, give rise to different trajectories.

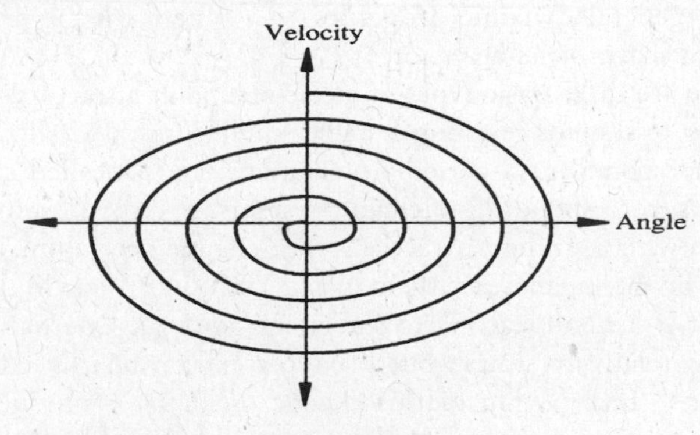

Figure 6-9
Phase space trajectory of a pendulum with friction.

Strange Attractors

Now let us return to our pendulum and notice that it was an idealized pendulum without friction, swinging back and forth in perpetual motion. This is a typical example of classical physics, where friction is generally neglected. A real pendulum will always have some friction that will slow it down so that, eventually, it will come to a halt. In the two-dimensional phase space this motion is represented by a curve spiraling inward toward the center, as shown in figure 6-9. This trajectory is called an "attractor," because mathematicians say, metaphorically, that the fixed point at the center of the coordinate system "attracts" the trajectory. The metaphor has been extended to include closed loops, such as the one representing the frictionless pendulum. A closed-loop trajectory is called a "periodic attractor," whereas the trajectory spiraling inward is called a "point attractor."

Over the past twenty years the phase-space technique has been used to explore a wide variety of complex systems. In case after case scientists and mathematicians would set up nonlinear equations, solve them numerically, and have computers trace out the solutions as trajectories in phase space. To their great surprise these researchers discovered that there is a very limited number of different attractors. Their shapes can be classified topologically, and the general dynamic properties of a system can be deduced from the shape of its attractor.

There are three basic types of attractors: point attractors, corresponding to systems reaching a stable equilibrium; periodic attractors, corresponding to periodic oscillations; and so-called strange attractors, corresponding to chaotic systems. A typical example of a system with a strange attractor is the "chaotic pendulum," studied first by the Japanese mathematician Yoshisuke Ueda in the late 1970s. It is a nonlinear electronic circuit with an external drive, which is relatively simple but produces extraordinarily complex behavior.[16] Each swing of this chaotic oscillator is unique. The system never repeats itself, so that each cycle covers a new region of phase space. However, in spite of the seemingly erratic motion, the points in phase space are not randomly distributed. Together they form a complex, highly organized pattern—a strange attractor, which now bears Ueda's name.

The Ueda attractor is a trajectory in a two-dimensional phase space that generates patterns that almost repeat themselves, but

Figure 6-10
The Ueda attractor; from Ueda et al. (1993).

not quite. This is a typical feature of all chaotic systems. The picture shown in figure 6-10 contains over one hundred thousand points. It may be visualized as a cut through a piece of dough that has been repeatedly stretched out and folded back on itself. Thus we see that the mathematics underlying the Ueda attractor is that of the "baker transformation."

One striking fact about strange attractors is that they tend to be of very low dimensionality, even in a high-dimensional phase space. For example, a system may have fifty variables, but its motion may be restricted to a strange attractor of three dimensions, a folded surface in that fifty-dimensional space. This, of course, represents a high degree of order.

Thus we see that chaotic behavior, in the new scientific sense of the term, is very different from random, erratic motion. With the help of strange attractors a distinction can be made between mere randomness, or "noise," and chaos. Chaotic behavior is deterministic and patterned, and strange attractors allow us to transform the seemingly random data into distinct visible shapes.

The "Butterfly Effect"

As we have seen in the case of the baker transformation, chaotic systems are characterized by extreme sensitivity to initial conditions. Minute changes in the system's initial state will lead over time to large-scale consequences. In chaos theory this is known as the "butterfly effect" because of the half-joking assertion that a butterfly stirring the air today in Beijing can cause a storm in New York next month. The butterfly effect was discovered in the early 1960s by the meteorologist Edward Lorenz, who designed a simple model of weather conditions consisting of three coupled nonlinear equations. He found that the solutions to his equations were extremely sensitive to the initial conditions. From virtually the same starting point, two trajectories would develop in completely different ways, making any long-range prediction impossible.[17]

This discovery sent shock waves through the scientific community, which was used to relying on deterministic equations for predicting phenomena such as solar eclipses or the appearance of comets with great precision over long spans of time. It seemed inconceivable that strictly deterministic equations of motion should lead to unpredictable results. Yet this was exactly what Lorenz had discovered. In his own words:

> The average person, seeing that we can predict tides pretty well a few months ahead, would say, why can't we do the same thing with the atmosphere, it's just a different fluid system, the laws are about as complicated. But I realized that *any* physical system that behaved nonperiodically would be unpredictable.[18]

The Lorenz model is not a realistic representation of a particular weather phenomenon, but it is a striking example of how a simple set of nonlinear equations can generate enormously complex behavior. Its publication in 1963 marked the beginning of chaos theory, and the model's attractor, known as the Lorenz attractor ever since, became the most celebrated and most widely studied strange attractor. Whereas the Ueda attractor lies in two dimensions, the Lorenz attractor is three-dimensional (figure 6-11). To trace it out, the point in phase space moves in an apparently random manner with a few oscillations of increasing amplitude around one point, followed by a few oscillations around a second point, then suddenly moving back again to oscillate around the first point, and so on.

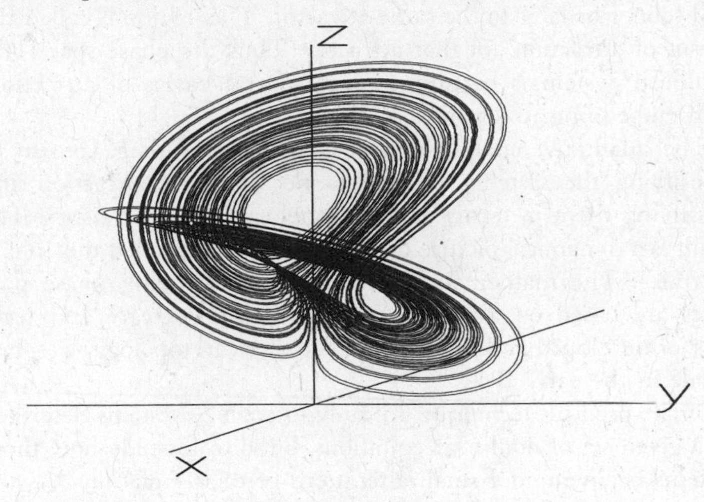

Figure 6-11
The Lorenz attractor; from Mosekilde et al. (1994).

From Quantity to Quality

The impossibility of predicting which point in phase space the trajectory of the Lorenz attractor will pass through at a certain time, even though the system is governed by deterministic equations, is a common feature of all chaotic systems. However, this does not mean that chaos theory is not capable of any predictions. We can still make very accurate predictions, but they concern the qualitative features of the system's behavior rather than the precise values of its variables at a particular time. The new mathematics thus represents a shift from quantity to quality that is characteristic of systems thinking in general. Whereas conventional mathematics deals with quantities and formulas, dynamical systems theory deals with quality and pattern.

Indeed, the analysis of nonlinear systems in terms of the topological features of their attractors is known as "qualitative analysis." A nonlinear system can have several attractors, which may be of different types, both "chaotic," or "strange," and nonchaotic. All trajectories starting within a certain region of phase space will lead sooner or later to the same attractor. This region is called the "basin of attraction" of that attractor. Thus the phase space of a nonlinear system is partitioned into several basins of attraction, each embedding its separate attractor.

The qualitative analysis of a dynamic system, then, consists in identifying the system's attractors and basins of attraction and classifying them in terms of their topological characteristics. The result is a dynamical picture of the entire system, called the "phase portrait." The mathematical methods for analyzing phase portraits are based on the pioneering work of Poincaré and were further developed and refined by the American topologist Stephen Smale in the early 1960s.[19]

Smale used his technique not only to analyze systems described by a given set of nonlinear equations, but also to study how those systems behave under small alterations of their equations. As the parameters of the equations change slowly, the phase portrait—for example, the shapes of its attractors and basins of attraction—will usually go through corresponding smooth alterations without any changes in its basic characteristics. Smale used the term "structurally stable" to describe such systems, in which small changes in the equations leave unchanged the basic character of the phase portrait.

In many nonlinear systems, however, small changes of certain parameters may produce dramatic changes in the basic characteristics of the phase portrait. Attractors may disappear or change into one another, or new attractors may suddenly appear. Such systems are said to be structurally unstable, and the critical points of instability are called "bifurcation points," because they are points in the system's evolution where a fork suddenly appears and the system branches off in a new direction. Mathematically bifurcation points mark sudden changes in the system's phase portrait. Physically they correspond to points of instability at which the system changes abruptly and new forms of order suddenly appear. As Prigogine has shown, such instabilities can occur only in open systems operating far from equilibrium.[20]

As there are only a small number of different types of attractors, so too are there only a small number of different types of bifurcation events; and like the attractors, the bifurcations can be classified topologically. One of the first to do so was the French mathematician René Thom in the 1970s, who used the term "catastrophes" instead of "bifurcations" and identified seven elementary catastrophes.[21] Today mathematicians know about three times as many bifurcation types. Ralph Abraham, professor of mathematics at the University of California at Santa Cruz, and graphic artist Christopher Shaw have created a series of visual mathematics books without any equations or formulas, which they see as the beginning of a complete encyclopedia of bifurcations.[22]

Fractal Geometry

While the first strange attractors were explored during the 1960s and 1970s, a new geometry, called "fractal geometry," was invented independently of chaos theory, which would provide a powerful mathematical language to describe the fine-scale structure of chaotic attractors. The author of this new language is the French mathematician Benoît Mandelbrot. In the late 1950s Mandelbrot began to study the geometry of a wide variety of irregular natural phenomena, and during the 1960s he realized that all these geometric forms had some very striking common features.

Over the next ten years Mandelbrot invented a new type of mathematics to describe and analyze these features. He coined the term "fractal" to characterize his invention and published his results in a spectacular book, *The Fractal Geometry of Nature,* which had a tremendous influence on the new generation of mathematicians who were developing chaos theory and other branches of dynamical systems theory.[23]

In a recent interview Mandelbrot explained that fractal geometry deals with an aspect of nature that almost everybody had been aware of but that nobody was able to describe in formal mathematical terms.[24] Some features of nature are geometric in the traditional sense. The trunk of a tree is more or less a cylinder; the full moon appears more or less as a circular disk; the planets go around the sun more or less in ellipses. But these are exceptions, Mandelbrot reminds us:

> Most of nature is very, very complicated. How could one describe a cloud? A cloud is not a sphere. . . . It is like a ball but very irregular. A mountain? A mountain is not a cone. . . . If you want to speak of clouds, of mountains, of rivers, of lightning, the geometric language of school is inadequate.

So Mandelbrot created fractal geometry—"a language to speak of clouds"—to describe and analyze the complexity of the irregular shapes in the natural world around us.

The most striking property of these "fractal" shapes is that their characteristic patterns are found repeatedly at descending scales, so that their parts, at any scale, are similar in shape to the whole. Mandelbrot illustrates this property of "self-similarity" by breaking a piece out of a cauliflower and pointing out that, by itself, the piece looks just like a small cauliflower.[25] He repeats this demonstration by dividing the part further, taking out another piece, which again looks like a very small cauliflower. Thus every part looks like the whole vegetable. The shape of the whole is similar to itself at all levels of scale.

There are many other examples of self-similarity in nature. Rocks on mountains look like small mountains; branches of lightning, or borders of clouds, repeat the same pattern again and again; coastlines divide into smaller and smaller portions, each showing similar arrangements of beaches and headlands. Photographs of a river delta, the ramifications of a tree, or the repeated branchings of blood vessels may show patterns of such striking similarity that we are unable to tell which is which. This similarity of images from vastly different scales has been known for a long time, but before Mandelbrot nobody had a mathematical language to describe it.

When Mandelbrot published his pioneering book in the mid-seventies, he was not aware of the connections between fractal geometry and chaos theory, but it did not take long for his fellow mathematicians and him to discover that strange attractors are exquisite examples of fractals. If parts of their structure are magnified, they reveal a multilayered substructure in which the same patterns are repeated again and again. Thus it has become customary to define strange attractors as trajectories in phase space that exhibit fractal geometry.

Another important link between chaos theory and fractal geometry is the shift from quantity to quality. As we have seen, it is impossible to predict the values of the variables of a chaotic system at a particular time, but we *can* predict the qualitative features of the system's behavior. Similarly, it is impossible to calculate the length or area of a fractal shape, but we can define the degree of "jaggedness" in a qualitative way.

Mandelbrot highlighted this dramatic feature of fractal shapes by asking a provocative question: How long is the coast of Britain? He showed that since the measured length can be extended indefinitely by going to smaller and smaller scales, there is no clear-cut answer to the question. However, it is possible to define a number between 1 and 2 that characterizes the jaggedness of the coast. For the British coastline this number is approximately 1.58; for the much rougher Norwegian coast it is approximately 1.70.[26]

Since it can be shown that this number has certain properties of a dimension, Mandelbrot called it a fractal dimension. We can understand this idea intuitively by realizing that a jagged line on a plane fills up more space than a smooth line, which has dimension 1, but less than the plane, which has dimension 2. The more jagged the line, the closer its fractal dimension will be to 2. Similarly, a crumpled-up piece of paper fills up more space than a plane but less than a sphere. Thus the more tightly the paper is crumpled, the closer its fractal dimension will be to 3.

This concept of a fractal dimension, which was at first a purely abstract mathematical idea, has become a very powerful tool for analyzing the complexity of fractal shapes, because it corresponds very well to our experience of nature. The more jagged the outlines of lightning or the borders of clouds, the rougher the shapes of coastlines or mountains, the higher their fractal dimensions.

To model the fractal shapes that occur in nature, geometric figures can be constructed that exhibit precise self-similarity. The principal technique for constructing these mathematical fractals is iteration—that is, repeating a certain geometric operation again and again. The process of iteration, which led us to the baker transformation, the mathematical characteristic underlying strange attractors, thus reveals itself as the central mathematical feature linking chaos theory and fractal geometry.

One of the simplest fractal shapes generated by iteration is the so-called Koch curve, or snowflake curve.[27] The geometric operation consists in dividing a line into three equal parts and replacing the center section by two sides of an equilateral triangle, as shown in figure 6-12. By repeating this operation again and again on smaller and smaller scales, a jagged snowflake is created (figure 6-13). Like a coastline, the Koch curve becomes infinitely long if the iteration is continued to infinity. Indeed, the Koch curve can be seen as a very rough model of a coastline (figure 6-14).

Figure 6-12
Geometric operation for constructing a Koch curve.

Figure 6-13
The Koch snowflake.

With the help of computers, simple geometric iterations can be applied thousands of times at different scales to produce so-called fractal forgeries—computer-generated models of plants, trees, mountains, coastlines, and so on that bear an astonishing resem-

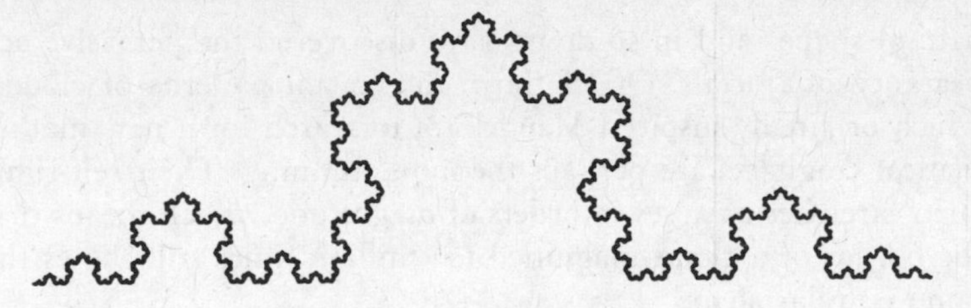

Figure 6-14
Modeling a coastline with the Koch curve.

blance to the actual shapes found in nature. Figure 6-15 shows an example of such a fractal forgery. By iterating a simple stick drawing at various scales, the beautiful and complex picture of a fern is generated.

Figure 6-15
Fractal forgery of a fern; from Garcia (1991).

With these new mathematical techniques scientists have been able to construct accurate models of a wide variety of irregular natural shapes and in so doing have discovered the pervasive appearance of fractals. Of all those, the fractal patterns of clouds, which originally inspired Mandelbrot to search for a new mathematical language, are perhaps the most stunning. Their self-similarity stretches over seven orders of magnitude, which means that the border of a cloud magnified ten million times still shows the same familiar shape.

Complex Numbers

The culmination of fractal geometry has been Mandelbrot's discovery of a mathematical structure that is of awesome complexity and yet can be generated with a very simple iterative procedure. To understand this amazing fractal figure, known as the Mandelbrot set, we need to first familiarize ourselves with one of the most important mathematical concepts—complex numbers.

The discovery of complex numbers is a fascinating chapter in the history of mathematics.[28] When algebra was developed in the Middle Ages and mathematicians explored all kinds of equations and classified their solutions, they soon came across problems that had no solution in terms of the set of numbers known to them. In particular, equations like $x + 5 = 3$ led them to extend the number concept to negative numbers, so that the solution could be written as $x = -2$. Later on, all so-called real numbers—positive and negative integers, fractions and irrational numbers (like square roots, or the famous number π)—were represented as points on a single, densely populated number line (figure 6-16).

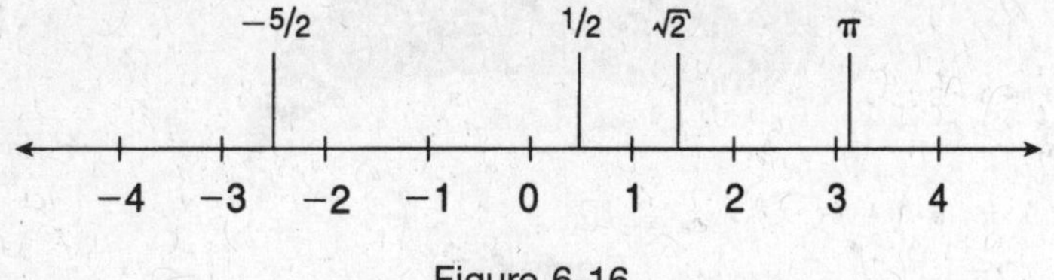

Figure 6-16
The number line.

With this expanded concept of numbers, all algebraic equations could be solved in principle except for those involving square roots of negative numbers. The equation $x^2 = 4$ has two solutions, $x = 2$ and $x = -2$; but for $x^2 = -4$ there seems to be no solution, because neither +2 nor −2 will give −4 when squared.

The early Indian and Arabic algebraists repeatedly encountered these equations, but they refused to write down expressions like $\sqrt{-4}$ because they thought them to be completely meaningless. It was not until the sixteenth century that square roots of negative numbers appeared in algebraic texts, and even then the authors were quick to point out that such expressions did not really mean anything.

Descartes called the square root of a negative number "imaginary" and believed that the occurrence of such "imaginary" numbers in a calculation meant that the problem had no solution. Other mathematicians used terms such as "fictitious," "sophisticated," or "impossible" to label those quantities that today, following Descartes, we still call "imaginary numbers."

Since the square root of a negative number cannot be placed anywhere on the number line, mathematicians up to the nineteenth century could not ascribe any sense of reality to those quantities. The great Leibniz, inventor of the differential calculus, attributed a mystical quality to the square root of −1, seeing it as a manifestation of "the Divine Spirit" and calling it "that amphibian between being and not-being."[29] A century later Leonhard Euler, the most prolific mathematician of all time, expressed the same sentiment in his *Algebra* in words that, even though less poetic, still echo the same sense of wonder:

> All such expressions as $\sqrt{-1}$, $\sqrt{-2}$, etc., are consequently impossible, or imaginary numbers, since they represent roots of negative quantities; and of such numbers we may truly assert that they are neither nothing, nor greater than nothing, nor less than nothing, which necessarily constitutes them imaginary or impossible.[30]

In the nineteenth century another mathematical giant, Karl Friedrich Gauss, finally declared forcefully that "an objective existence can be assigned to these imaginary beings."[31] Gauss realized, of course, that there was no room for imaginary numbers anywhere on the number line, so he took the bold step of placing them on a perpendicular axis through the point zero, thus creating a Cartesian coordinate system. In this system all real numbers are placed on the "real axis" and all imaginary numbers on the "imaginary axis" (figure 6-17). The square root of -1 is called the "imaginary unit" and given the symbol *i,* and since any square root of a negative number can always be written as $\sqrt{-a} = \sqrt{-1}\sqrt{a} = i\sqrt{a}$, all imaginary numbers can be placed on the imaginary axis as multiples of *i.*

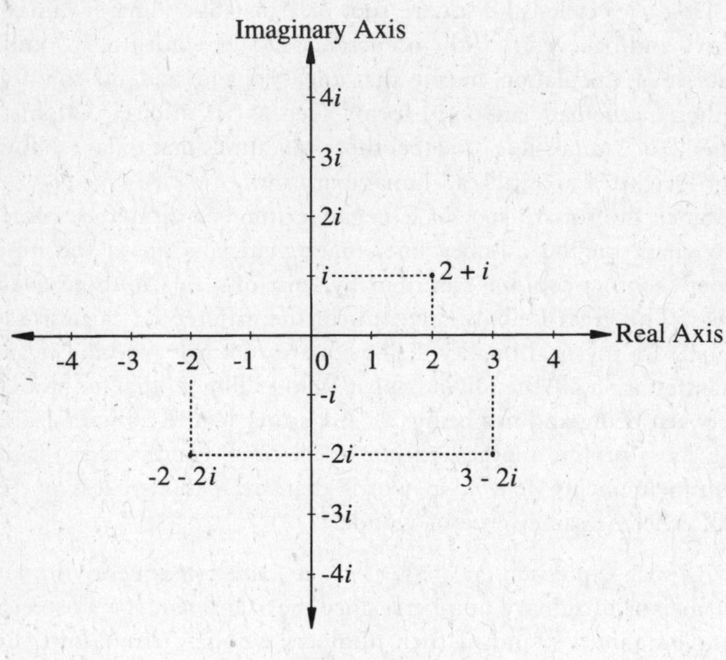

Figure 6-17
The complex plane.

With this ingenious device Gauss created a home not only for imaginary numbers, but also for all possible combinations of real and imaginary numbers, such as $(2 + i)$, $(3 - 2i)$, and so on. Such combinations are called "complex numbers" and are represented by points in the plane spanned by the real and imaginary axes, which is called the "complex plane." In general, any complex number can be written as

$$z = x + iy$$

where x is called the "real part" and y the "imaginary part."

With the help of this definition Gauss created a special algebra of complex numbers and developed many fundamental ideas about functions of complex variables. Eventually this led to a whole new branch of mathematics, known as "complex analysis," which has an enormous range of applications in all fields of science.

Patterns within Patterns

The reason why we took this excursion into the history of complex numbers is that many fractal shapes can be generated mathematically by iterative procedures in the complex plane. In the late seventies, after publishing his pioneering book, Mandelbrot turned his attention to a particular class of those mathematical fractals known as Julia sets.[32] They had been discovered by the French mathematician Gaston Julia during the early part of the century but had soon faded into obscurity. In fact, Mandelbrot had come across Julia's work as a student, had looked at his primitive drawings (done at that time without the help of a computer), and had soon lost interest. Now, however, Mandelbrot realized that Julia's drawings were rough renderings of complex fractal shapes, and he proceeded to reproduce them in fine detail with the most powerful computers he could find. The results were stunning.

The basis of the Julia set is the simple mapping

$$z \rightarrow z^2 + c$$

where z is a complex variable and c a complex constant. The iterative procedure consists in picking any number z in the complex plane, squaring it, adding the constant c, squaring the result again, adding the constant c once more, and so on. When this is done with different starting values for z, some of them will keep increasing and move to infinity as the iteration proceeds, while others will remain finite.[33] The Julia set is the set of all those values of z, or points in the complex plane, that remain finite under the iteration.

To determine the shape of the Julia set for a particular constant c, the iteration has to be carried out for thousands of points, each time until it becomes clear whether they will keep increasing or remain finite. If those points that remain finite are colored black, while those that keep increasing remain white, the Julia set will emerge as a black shape in the end. The entire procedure is very simple but very time-consuming. It is evident that the use of a high-speed computer is essential if one wants to obtain a precise shape in a reasonable time.

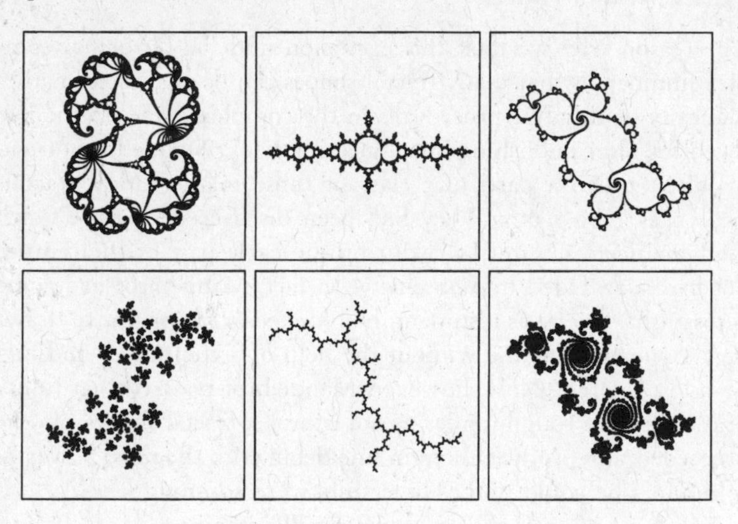

Figure 6-18
Varieties of Julia sets; from Peitgen and Richter (1986).

For each constant c one will obtain a different Julia set, so there is an infinite number of these sets. Some are single connected pieces; others are broken into several disconnected parts; yet others look as though they have burst into dust (figure 6-18). All have the jagged look that is characteristic of fractals, and most of them are impossible to describe in the language of classical geometry. "You obtain an incredible variety of Julia sets," marvels French mathematician Adrien Douady. "Some are a fatty cloud, others are a skinny bush of brambles, some look like the sparks which float in the air after a firework has gone off. One has the shape of a rabbit, lots of them have seahorse tails."[34]

This rich variety of forms, many of which are reminiscent of living things, is amazing enough. But the real magic begins when we magnify the contour of any portion of a Julia set. As in the case of a cloud or coastline, the same richness is displayed across all scales. With increasing resolution (that is, with more and more decimals of the number z entering into the calculation) more and more details of the fractal contour appear, revealing a fantastic sequence of patterns within patterns—all similar without ever being identical.

When Mandelbrot analyzed different mathematical representations of Julia sets in the late seventies and tried to classify their immense variety, he discovered a very simple way of creating a single image in the complex plane that would serve as a catalog of all possible Julia sets. That image, which has since become the principal visual symbol of the new mathematics of complexity, is the Mandelbrot set (figure 6-19). It is simply the collection of all points of the constant c in the complex plane for which the corresponding Julia sets are single connected pieces. To construct the Mandelbrot set, therefore, one needs to construct a separate Julia set for each point c in the complex plane and determine whether that particular Julia set is "connected" or "disconnected." For example, among the Julia sets shown in figure 6-18, the three sets in the top row and the one in the center panel of the bottom row are connected (that is, they consist of a single piece), while the two sets in the side panels of the bottom row are disconnected (consist of several pieces).

To generate Julia sets for thousands of values of c, each involving thousands of points requiring repeated iterations, seems an impossible task. Fortunately, however, there is a powerful theo-

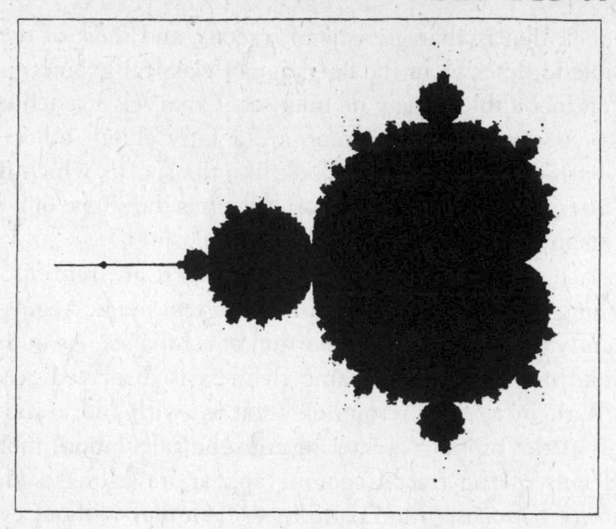

Figure 6-19
The Mandelbrot set; from Peitgen and Richter (1986).

rem, discovered by Gaston Julia himself, which drastically reduces the number of necessary steps.[35] To find out whether a particular Julia set is connected or disconnected, all one has to do is iterate the starting point $z = 0$. If that point remains finite under repeated iteration, the Julia set is always connected, however crumpled it may be; if not, it is always disconnected. Therefore one really needs to iterate only that one point, $z = 0$, for each value of c to construct the Mandelbrot set. In other words, generating the Mandelbrot set involves the same number of steps as generating a Julia set.

While there is an infinite number of Julia sets, the Mandelbrot set is unique. This strange figure is the most complex mathematical object ever invented. Although the rules for its construction are very simple, the variety and complexity it reveals upon close inspection is unbelievable. When the Mandelbrot set is generated on a rough grid, two disks appear on the computer screen: the smaller one approximately circular, the larger one vaguely heart shaped. Each of the two disks shows several smaller disklike attachments to its boundary, and further resolution reveals a profusion of smaller and smaller attachments looking not unlike prickly thorns.

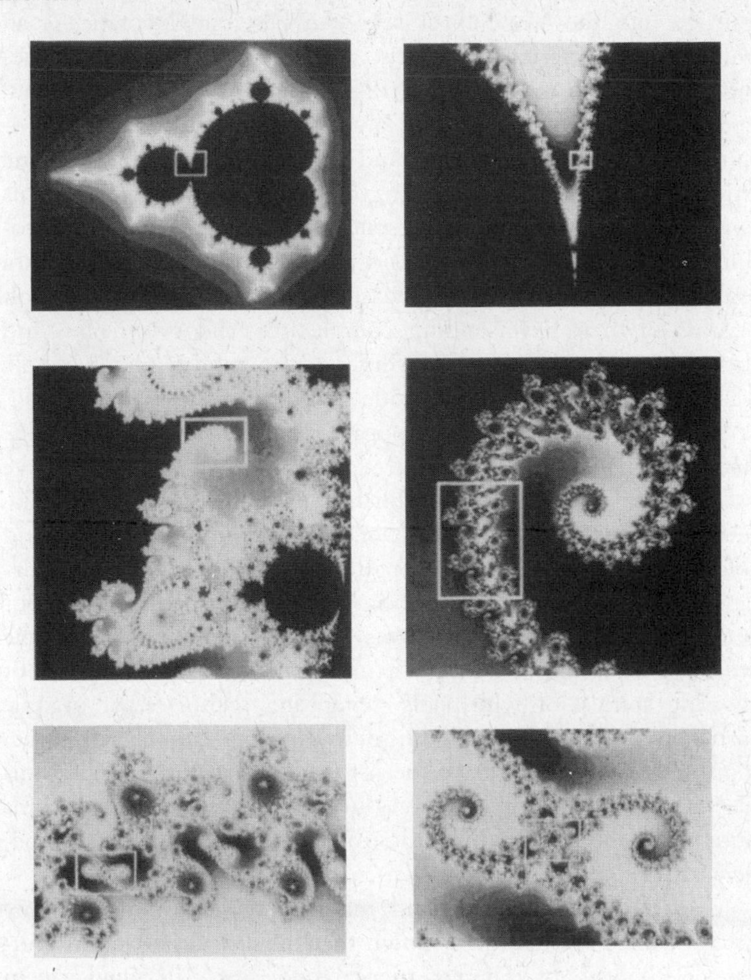

Figure 6-20
Stages of a journey into the Mandelbrot set. In each picture the area of the Subsequent magnification is marked with a white rectangle; from Peitgen and Richter (1986).

From this point on, the wealth of images revealed by increasing magnification of the set's boundary (that is, by increasing resolution in the calculations) is almost impossible to describe. Such a journey into the Mandelbrot set, seen best on videotape, is an unforgettable experience.[36] As the camera zooms in and magnifies the boundary, sprouts and tendrils seem to grow out from it that, upon further magnification, dissolve into a multitude of shapes—spirals within spirals, seahorses and whirlpools, repeating the same patterns over and over again (figure 6-20). At each scale of this fantastic journey—in which present-day computer power can produce magnifications up to a hundred million times!—the picture looks like a richly fragmented coast, but featuring forms that look organic in their never-ending complexity. And every now and then we make an eerie discovery—a tiny replica of the whole Mandelbrot set buried deep inside its boundary structure.

Since the Mandelbrot set appeared on the cover of *Scientific American* in August 1985, hundreds of computer enthusiasts have used the iterative program published in that issue to undertake their own journeys into the set on their home computers. Vivid colors have been added to the patterns discovered on those journeys, and the resulting pictures have been published in numerous books and shown in exhibitions of computer art around the world.[37] Looking at these hauntingly beautiful pictures of swirling spirals, of whirlpools generating seahorses, of organic forms burgeoning and exploding into dust, one cannot help noticing the striking similarity to the psychedelic art of the 1960s. This was an art inspired by similar journeys, facilitated not by computers and the new mathematics, but by LSD and other psychedelic drugs.

The term *psychedelic* ("mind manifesting") was invented because detailed research had shown that these drugs act as amplifiers, or catalysts, of inherent mental processes.[38] It would seem therefore that the fractal patterns that are such a striking characteristic of the LSD experience must, somehow, be embedded in the human brain. The fact that fractal geometry and LSD appeared on the scene at roughly the same time is one of those amazing coincidences—or synchronicities?—that have occurred so often in the history of ideas.

The Mandelbrot set is a storehouse of patterns of infinite detail and variations. Strictly speaking, it is not self-similar because it not only repeats the same patterns over and over again, including small replicas of the entire set, but also contains elements from an infinite number of Julia sets! It is thus a "superfractal" of inconceivable complexity.

Yet this structure whose richness defies the human imagination is generated by a few very simple rules. Thus fractal geometry, like chaos theory, has forced scientists and mathematicians to re-examine the very concept of complexity. In classical mathematics simple formulas correspond to simple shapes, complicated formulas to complicated shapes. In the new mathematics of complexity the situation is dramatically different. Simple equations may generate enormously complex strange attractors, and simple rules of iteration give rise to structures more complicated than we can even imagine. Mandelbrot sees this as a very exciting new development in science:

> It's a very optimistic conclusion because, after all, the initial meaning of the study of chaos was the attempt to find simple rules in the universe around us. . . . The effort was always to seek simple explanations for complicated realities. But the discrepancy between simplicity and complexity was never anywhere comparable to what we find in this context.[39]

Mandelbrot also sees the tremendous interest in fractal geometry outside the mathematics community as a healthy development. He hopes that it will end the isolation of mathematics from other human activities and the consequent widespread ignorance of mathematical language even among otherwise highly educated people.

This isolation of mathematics is a striking sign of our intellectual fragmentation and as such is a relatively recent phenomenon. Throughout the centuries many of the great mathematicians made outstanding contributions to other fields as well. In the eleventh century the Persian poet Omar Khayyám, who is world renowned as the author of the *Rubáiyát,* also wrote a pioneering book on algebra and served as the official astronomer at the caliph's court. Descartes, the founder of modern philosophy, was a brilliant mathematician and also practiced medicine. Both inventors of the differential calculus, Newton and Leibniz, were active in many fields besides mathematics. Newton was a "natural philosopher" who made fundamental contributions to virtually all branches of science that were known at his time, in addition to studying alchemy, theology, and history. Leibniz is known primarily as a philosopher, but he was also the founder of symbolic logic and was active as a diplomat and historian during most of his life. The great mathematician Gauss was also a physicist and astronomer, and he invented several useful instruments, including the electric telegraph.

These examples, to which dozens more could be added, show that throughout our intellectual history mathematics was never separated from other areas of human knowledge and activity. In the twentieth century, however, increasing reductionism, fragmentation, and specialization led to an extreme isolation of mathematics, even within the scientific community. Thus chaos theorist Ralph Abraham remembers:

> When I started my professional work in mathematics in 1960, which is not so long ago, modern mathematics in its entirety—in its entirety—was rejected by physicists, including the most avant-garde mathematical physicists. . . . Everything just a year or two beyond what Einstein had used was all rejected. . . . Mathematical physicists refused their graduate students permission to take math courses from mathematicians: "Take mathematics from us. We will teach you what you need to know. . . ." That was in 1960. By 1968 this had completely turned around.[40]

The great fascination exerted by chaos theory and fractal geometry on people in all disciplines—from scientists to managers to artists—may indeed be a hopeful sign that the isolation of mathematics is ending. Today the new mathematics of complexity is making more and more people realize that mathematics is much more than dry formulas; that the understanding of pattern is crucial to understand the living world around us; and that all questions of pattern, order, and complexity are essentially mathematical.

PART FOUR

The Nature of Life

7

A New Synthesis

We can now return to the central question of this book: What is life? My thesis has been that a theory of living systems consistent with the philosophical framework of deep ecology, including an appropriate mathematical language and implying a nonmechanistic, post-Cartesian understanding of life, is now emerging.

Pattern and Structure

The emergence and refinement of the concept of "pattern of organization" has been a crucial element in the development of this new way of thinking. From Pythagoras to Aristotle, to Goethe, and to the organismic biologists, there is a continuous intellectual tradition that struggles with the understanding of pattern, realizing that it is crucial to the understanding of living form. Alexander Bogdanov was the first to attempt the integration of the concepts of organization, pattern, and complexity into a coherent systems theory. The cyberneticists focused on patterns of communication and control—in particular on the patterns of circular causality underlying the feedback concept—and in doing so were the first to clearly distinguish the pattern of organization of a system from its physical structure.

The missing "pieces of the puzzle" were identified and analyzed over the past twenty years—the concept of self-organization and the new mathematics of complexity. Again the notion of pattern has been central to both of these developments. The concept of self-organization originated in the recognition of the network as the general pattern of life, which was subsequently refined by Maturana and Varela in their concept of autopoiesis. The new mathematics of complexity is essentially a mathematics of visual patterns—strange attractors, phase portraits, fractals, and so on—which are analyzed within the framework of topology pioneered by Poincaré.

The understanding of pattern, then, will be of crucial importance to the scientific understanding of life. However, for a full understanding of a living system, the understanding of its pattern of organization, although critically important, is not enough. We also need to understand the system's structure. Indeed, we have seen that the study of structure has been the principal approach in Western science and philosophy and as such has again and again eclipsed the study of pattern.

I have come to believe that the key to a comprehensive theory of living systems lies in the synthesis of those two approaches—the study of pattern (or form, order, quality) and the study of structure (or substance, matter, quantity). I shall follow Humberto Maturana and Francisco Varela in their definitions of those two key criteria of a living system—its pattern of organization and its structure.[1] The *pattern of organization* of any system, living or nonliving, is the configuration of relationships among the system's components that determines the system's essential characteristics. In other words, certain relationships must be present for something to be recognized as—say—a chair, a bicycle, or a tree. That configuration of relationships that gives a system its essential characteristics is what we mean by its pattern of organization.

The *structure* of a system is the physical embodiment of its pattern of organization. Whereas the description of the pattern of organization involves an abstract mapping of relationships, the description of the structure involves describing the system's actual physical components—their shapes, chemical compositions, and so forth.

To illustrate the difference between pattern and structure, let us look at a well-known nonliving system, a bicycle. In order for something to be called a bicycle, there must be a number of functional relationships among components known as frame, pedals, handlebars, wheels, chain, sprocket, and so on. The complete configuration of these functional relationships constitutes the bicycle's pattern of organization. All of those relationships must be present to give the system the essential characteristics of a bicycle.

The structure of the bicycle is the physical embodiment of its pattern of organization in terms of components of specific shapes, made of specific materials. The same pattern "bicycle" can be embodied in many different structures. The handlebars will be shaped differently for a touring bike, a racing bike, or a mountain bike; the frame may be heavy and solid or light and delicate; the tires may be narrow or wide, tubes or solid rubber. All these combinations and many more will easily be recognized as different embodiments of the same pattern of relationships that defines a bicycle.

The Three Key Criteria

In a machine such as a bicycle the parts have been designed, manufactured, and then put together to form a structure with fixed components. In a living system, by contrast, the components change continually. There is a ceaseless flux of matter through a living organism. Each cell continually synthesizes and dissolves structures and eliminates waste products. Tissues and organs replace their cells in continual cycles. There is growth, development, and evolution. Thus from the very beginning of biology, the understanding of living structure has been inseparable from the understanding of metabolic and developmental processes.[2]

This striking property of living systems suggests *process* as a third criterion for a comprehensive description of the nature of life. The process of life is the activity involved in the continual embodiment of the system's pattern of organization. Thus the process criterion is the link between pattern and structure. In the case of the bicycle, the pattern of organization is represented by the design sketches that are used to build the bicycle, the structure is a specific physical bicycle, and the link between pattern and structure is in the mind of the designer. In the case of a living organism, however, the pattern of organization is always embodied in the organism's structure, and the link between pattern and structure lies in the process of continual embodiment.

The process criterion completes the conceptual framework of my synthesis of the emerging theory of living systems. The definitions of the three criteria—pattern, structure, and process—are listed once more in the table that follows. All three criteria are totally interdependent. The pattern of organization can be recognized only if it is embodied in a physical structure, and in living systems this embodiment is an ongoing process. Thus structure and process are inextricably linked. One could say that the three criteria—pattern, structure, and process—are three different but inseparable perspectives on the phenomenon of life. They will form the three conceptual dimensions of my synthesis.

To understand the nature of life from a systemic point of view means to identify a set of general criteria by which we can make a clear distinction between living and nonliving systems. Throughout the history of biology many criteria have been suggested, but all of them turned out to be flawed in one way or another. However, the recent formulations of models of self-organization and the mathematics of complexity indicate that it is now possible to identify such criteria. The key idea of my synthesis is to express those criteria in terms of the three conceptual dimensions, pattern, structure, and process.

In a nutshell, I propose to understand autopoiesis, as defined by Maturana and Varela, as the pattern of life (that is, the pattern of organization of living systems);[3] dissipative structure, as defined by Prigogine, as the structure of living systems;[4] and cognition, as defined initially by Gregory Bateson and more fully by Maturana and Varela, as the process of life.

The pattern of organization determines a system's essential characteristics. In particular it determines whether the system is

Key Criteria of a Living System

pattern of organization
the configuration of relationships that determines the system's essential characteristics

structure
the physical embodiment of the system's pattern of organization

life process
the activity involved in the continual embodiment of the system's pattern of organization

---

living or nonliving. Autopoiesis—the pattern of organization of living systems—is thus the defining characteristic of life in the new theory. To find out whether a particular system—a crystal, a virus, a cell, or the planet Earth—is alive, all we need to do is find out whether its pattern of organization is that of an autopoietic network. If it is, we are dealing with a living system; if it is not, the system is nonliving.

Cognition, the process of life, is inextricably linked to autopoiesis, as we shall see. Autopoiesis and cognition are two different aspects of the same phenomenon of life. In the new theory all living systems are cognitive systems, and cognition always implies the existence of an autopoietic network.

With the third criterion of life, the structure of living systems, the situation is slightly different. Although the structure of a living system is always a dissipative structure, not all dissipative structures are autopoietic networks. Thus a dissipative structure may be a living or a nonliving system. For example, the Bénard cells and chemical clocks studied extensively by Prigogine are dissipative structures but not living systems.[5]

The three key criteria of life and the theories underlying them will be discussed in detail in the following chapters. At this point I merely want to give a brief overview.

Autopoiesis—the Pattern of Life

Since the early part of the century it has been known that the pattern of organization of a living system is always a network pattern.[6] However, we also know that not all networks are living systems. According to Maturana and Varela, the key characteristic of a living network is that it continually produces itself. Thus "the being and doing of [living systems] are inseparable, and this is their specific mode of organization."[7] Autopoiesis, or "self-making," is a network pattern in which the function of each component is to participate in the production or transformation of other components in the network. In this way the network continually makes itself. It is produced by its components and in turn produces those components.

The simplest living system we know is a cell, and Maturana and Varela have used cell biology extensively to explore the details of autopoietic networks. The basic pattern of autopoiesis can be illustrated conveniently with a plant cell. Figure 7-1 shows a simplified picture of such a cell, in which the components have been given descriptive English names. The corresponding technical terms, derived from Greek and Latin, are listed in the glossary that follows.

Like every other cell, a typical plant cell consists of a cell membrane which encloses the cell fluid. The fluid is a rich molecular soup of cell nutrients—that is, of the chemical elements out of which the cell builds its structures. Suspended in the cell fluid we find the cell nucleus, a large number of tiny production centers where the main structural building blocks are produced, and several specialized parts, called "organelles," which are analogous to body organs. The most important of these organelles are the storage sacs, recycling centers, powerhouses, and solar stations. Like the cell as a whole, the nucleus and the organelles are surrounded by semipermeable membranes that select what comes in and what goes out. The cell membrane, in particular, takes in food and dissipates waste.

The cell nucleus contains the genetic material—the DNA mole-

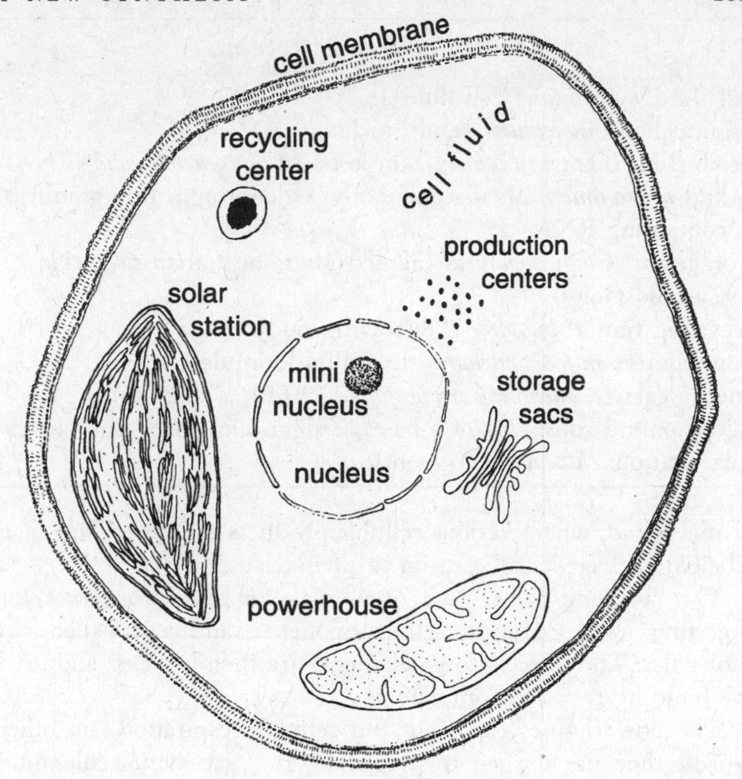

Figure 7-1
Basic components of a plant cell.

cules carrying the genetic information, and the RNA molecules, which are made by the DNA to deliver instructions to the production centers.[8] The nucleus also contains a smaller "mininucleus," where the production centers are made before being distributed throughout the cell.

The production centers are granular bodies in which the cell's proteins are produced. These include structural proteins as well as the enzymes, the catalysts that promote all cellular processes. There are about five hundred thousand production centers in each cell.

The storage sacs are stacks of flat pouches, somewhat like a pile

Glossary of Technical Terms

cell fluid: *cytoplasm* ("cell fluid")
mininucleus: *nucleolus* ("small nucleus")
production center: *ribosome;* composite of *ribonucleic acid* (RNA) and *microsome* ("microscopic body"), denoting a tiny granule containing RNA
storage sac: *Golgi apparatus* (named after the Italian physician Camillo Golgi)
recycling center: *lysosome* ("dissolving body")
powerhouse: *mitochondrion* ("threadlike granule")
energy carrier: *adenosine triphosphate* (ATP), a chemical compound consisting of a base, a sugar, and three phosphates
solar station: *chloroplast* ("green leaf")

---

of pita bread, where various cellular products are stored and then labeled, packaged, and sent on to their destinations.

The recycling centers are organelles containing enzymes for digesting food, damaged cell components, and various unused molecules. The broken-down elements are then recycled and used for building new cell components.

The powerhouses carry out the cellular respiration—in other words, they use oxygen to break down organic molecules into carbon dioxide and water. This releases energy that is locked up in special energy carriers. These energy carriers are complex molecular compounds that travel to the other parts of the cell to supply energy for all cellular processes, known collectively as "cell metabolism." The energy carriers serve as the cell's main energy units, not unlike cash in the human economy.

It was discovered only recently that the powerhouses contain their own genetic material and replicate independently of the replication of the cell. According to a theory by Lynn Margulis, they evolved from simple bacteria that came to live in the complex larger cells about two billion years ago.[9] Since then they have been permanent residents in all higher organisms, passed on from generation to generation and living in intimate symbiosis with each cell.

Like the powerhouses, the solar stations contain their own genetic material and self-reproduce, but they are found only in green plants. They are the centers for photosynthesis, transforming solar energy, carbon dioxide, and water into sugars and oxygen. The sugars then travel to the powerhouses, where their energy is extracted and stored in energy carriers. To supplement the sugars, plants also absorb nutrients and trace elements from the earth through their roots.

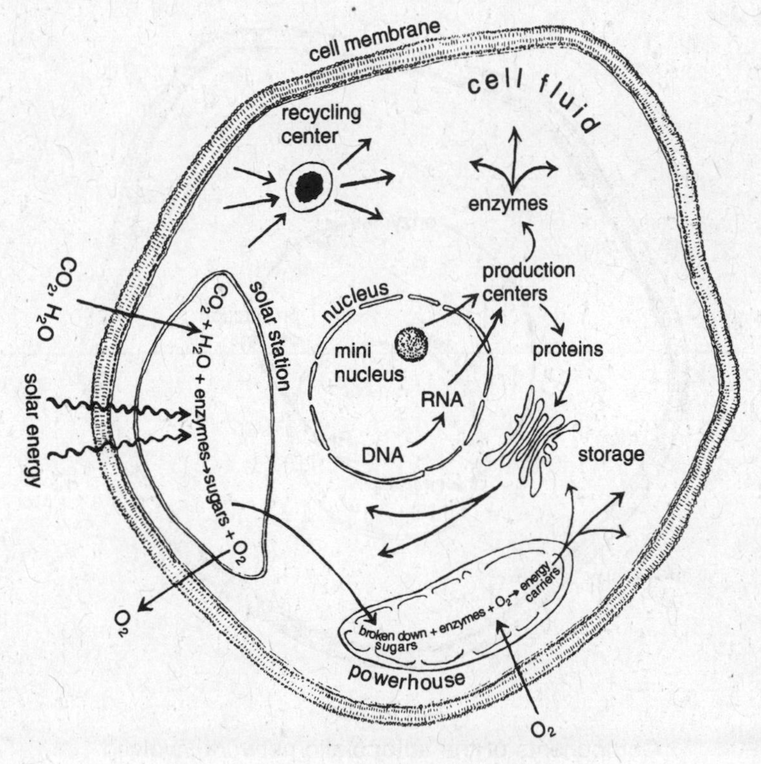

Figure 7-2
Metabolic processes in a plant cell.

We see that in order to give even a rough idea of cellular organization, the description of the cell's components has to be quite elaborate; and the complexity increases dramatically when we try to picture how these cell components are interlinked in a vast network, involving thousands of metabolic processes. The enzymes alone form an intricate network of catalytic reactions, promoting all metabolic processes, and the energy carriers form a corresponding energy network to fuel them. Figure 7-2 shows another drawing of our simplified plant cell, this time with various arrows indicating some of the links in the network of metabolic processes.

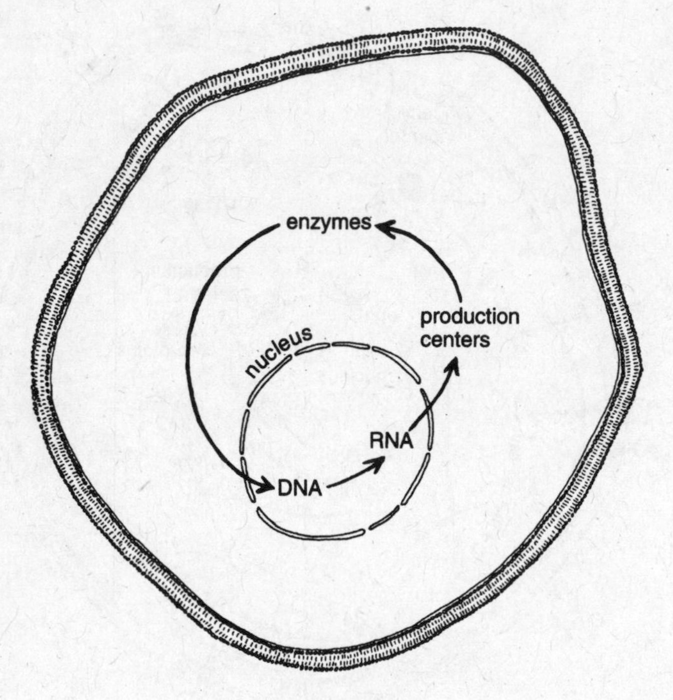

Figure 7-3
Components of the autopoietic network involved in the repair of DNA.

To illustrate the nature of this network, let us look at just one single loop. The DNA in the cell nucleus produces RNA molecules, which contain instructions for the production of proteins, including enzymes. Among these is a group of special enzymes that can recognize, remove, and replace damaged sections of DNA.[10] Figure 7-3 is a schematic drawing of some of the relationships involved in this loop. The DNA produces RNA, which delivers instructions to the production centers for producing the enzymes, which enter the cell nucleus to repair the DNA. Each component in this partial network helps to produce or transform other components; thus the network is clearly autopoietic. The DNA produces the RNA; the RNA specifies the enzymes; and the enzymes repair the DNA.

To complete the picture, we would have to add the building blocks from which DNA, RNA, and enzymes are made; the energy carriers fueling each of the processes pictured; the generation of the energy in the powerhouses from broken-down sugars; the production of the sugars by photosynthesis in the solar stations; and so on. With each addition to the network we would see that the new components, too, help to produce and transform other components, and thus the autopoietic, self-making nature of the entire network would become ever more apparent.

The case of the cell membrane is especially interesting. It is a boundary of the cell, formed by some of the cell's components, which encloses the network of metabolic processes and thus limits their extension. At the same time, the membrane participates in the network by selecting the raw material for the production processes (the cell's food) through special filters and by dissipating waste into the outside environment. Thus the autopoietic network creates its own boundary, which defines the cell as a distinct system while being an active part of the network.

Since all components of an autopoietic network are produced by other components in the network, the entire system is *organizationally closed,* even though it is open with regard to the flow of energy and matter. This organizational closure implies that a living system is self-organizing in the sense that its order and behavior are not imposed by the environment but are established by the system itself. In other words, living systems are autonomous. This does not mean that they are isolated from their environment. On the contrary, they interact with the environment through a continual exchange of energy and matter. But this interaction does not determine their organization—they are *self*-organizing. Autopoiesis, then, is seen as the pattern underlying the phenomenon of self-organization, or autonomy, that is so characteristic of all living systems.

Through their interactions with the environment living organisms continually maintain and renew themselves, using energy and resources from the environment for that purpose. Moreover, the continual self-making also includes the ability to form new structures and new patterns of behavior. We shall see that this creation of novelty, resulting in development and evolution, is an intrinsic aspect of autopoiesis.

A subtle but important point in the definition of autopoiesis is the fact that an autopoietic network is not a set of relations among static *components* (like, for example, the pattern of organization of a crystal), but a set of relations among *processes of production* of components. If these processes stop, so does the entire organization. In other words, autopoietic networks must continually regenerate themselves to maintain their organization. This, of course, is a well-known characteristic of life.

Maturana and Varela see the difference between relationships among static components and relationships among processes as a key distinction between physical and biological phenomena. Since the processes in a biological phenomenon involve components, it is always possible to abstract from them a description of those components in purely physical terms. However, the authors argue that such a purely physical description will not capture the biological phenomenon. A biological explanation, they maintain, must be one in terms of relationships of processes within the context of autopoiesis.

Dissipative Structure—the Structure of Living Systems

When Maturana and Varela describe the pattern of life as an autopoietic network, their main emphasis is on the organizational closure of that pattern. When Ilya Prigogine describes the structure of a living system as a dissipative structure, by contrast, his main emphasis is on the openness of that structure to the flow of energy and matter. Thus a living system is both open and closed—it is structurally open, but organizationally closed. Matter continually flows through it, but the system maintains a stable form, and it does so autonomously through self-organization.

Figure 7-4
Vortex funnel of whirlpool in a bathtub.

To highlight that seemingly paradoxical coexistence of change and stability, Prigogine coined the term "dissipative structures." As I've already mentioned, not all dissipative structures are living systems, and to visualize the coexistence of continual flow and structural stability, it is easier to turn to simple, nonliving dissipative structures. One of the simplest structures of this kind is a vortex in flowing water—for example, a whirlpool in a bathtub. Water continuously flows through the vortex, yet its characteristic shape, the well-known spirals and narrowing funnel, remains remarkably stable (figure 7-4). It is a dissipative structure.

Closer examination of the origin and progression of such a vortex reveals a series of rather complex phenomena.[11] Imagine a bathtub with shallow, motionless water. When the drain is opened, the water begins to run out, flowing radially toward the drain and speeding up as it approaches the hole under the accelerating force of gravity. Thus a smooth uniform flow is established. The flow does not remain in this smooth state for long, however.

Tiny irregularities in the water movement, movements of the air at the water's surface, and irregularities in the drainpipe will cause a little more water to approach the drain on one side than on the other, and thus a whirling, rotary motion is introduced into the flow.

As the water particles are dragged down toward the drain, both their radial and rotational velocities increase. They speed up radially because of the accelerating force of gravity, and they pick up rotational speed as the radius of their rotation decreases, like a skater pulling in her arms during a pirouette.[12] As a result, the water particles move downward in spirals, forming a narrowing tube of flow lines, known as a vortex tube.

Because the basic flow is still radially inward, the vortex tube is continually squeezed by the water pressing against it from all sides. This pressure decreases its radius and intensifies the rotation further. Using Prigogine's language, we can say that the rotation introduces an instability into the initial uniform flow. The force of gravity, the water pressure, and the constantly diminishing radius of the vortex tube all combine to accelerate the whirling motion to ever-increasing speeds.

However, this continuing acceleration ends not in catastrophe but in a new stable state. At a certain rotational speed, centrifugal forces come into play that push the water radially away from the drain. Thus the water surface above the drain develops a depression, which quickly turns into a funnel. Eventually a miniature tornado of air forms inside this funnel, creating highly complex and nonlinear structures—ripples, waves, and eddies—on the water surface inside the vortex.

In the end the force of gravity pulling the water down the drain, the water pressure pushing inward, and the centrifugal forces pushing outward balance each other and result in a stable state, in which gravity maintains the flow of energy at the larger scale, and friction dissipates some of it at smaller scales. The acting forces are now interlinked in self-balancing feedback loops that give great stability to the vortex structure as a whole.

Similar dissipative structures of great stability arise in thunderstorms under special atmospheric conditions. Hurricanes and tornadoes are vortices of violently rotating air, which can travel over large distances and unleash destructive forces without significant changes in their vortex structure. The detailed phenomena in these atmospheric vortices are much richer than those in the bathtub whirlpool, because several new factors come into play—temperature differences, expansions and contractions of air, moisture effects, condensations and evaporations, and so forth. The resulting structures are thus much more complex than the whirlpools in flowing water and display a greater variety of dynamic behaviors. Thunderstorms can turn into dissipative structures with characteristic sizes and shapes; under special conditions some of them can even split in two.

Metaphorically we can also visualize a cell as a whirlpool—that is, as a stable structure with matter and energy continually flowing through it. However, the forces and processes at work in a cell are quite different—and vastly more complex—than those in a vortex. While the balancing forces in the whirlpool are mechanical, the dominant force being gravity, those in the cell are chemical. More precisely they are the catalytic loops in the cell's autopoietic network that act as self-balancing feedback loops.

Similarly, the origin of the whirlpool's instability is mechanical, arising as a consequence of the first rotary motion. In a cell there are different kinds of instabilities, and their nature is chemical rather than mechanical. They too originate in the catalytic cycles that are a central feature of all metabolic processes. The crucial property of these cycles is their ability to act not only as self-balancing but also as self-amplifying feedback loops, which may push the system farther and farther away from equilibrium until it reaches a threshold of stability. This point is called a "bifurcation point." It is a point of instability at which new forms of order may emerge spontaneously, resulting in development and evolution.

Mathematically a bifurcation point represents a dramatic change of the system's trajectory in phase space.[13] A new attractor may suddenly appear, so that the system's behavior as a whole "bifurcates," or branches off, in a new direction. Prigogine's detailed studies of these bifurcation points have revealed some fascinating properties of dissipative structures, as we shall see in a subsequent chapter.[14]

The dissipative structures formed by whirlpools or hurricanes can maintain their stability only as long as there is a steady flow of matter from the environment through the structure. Similarly, a living dissipative structure, such as an organism, needs a continual flow of air, water, and food from the environment through the system in order to stay alive and maintain its order. The vast network of metabolic processes keeps the system in a state far from equilibrium and, through its inherent feedback loops, gives rise to bifurcations and thus to development and evolution.

Cognition—the Process of Life

The three key criteria of life—pattern, structure, and process—are so closely intertwined that it is difficult to discuss them separately, although it is important to distinguish among them. Autopoiesis, the pattern of life, is a set of relationships among *processes* of production; and a dissipative structure can be understood only in terms of metabolic and developmental *processes.* The process dimension is thus implicit both in the pattern and in the structure criterion.

In the emerging theory of living systems the process of life—the continual embodiment of an autopoietic pattern of organization in a dissipative structure—is identified with cognition, the process of knowing. This implies a radically new concept of mind, which is perhaps the most revolutionary and most exciting aspect of this theory, as it promises finally to overcome the Cartesian division between mind and matter.

According to the theory of living systems, mind is not a thing but a process—the very process of life. In other words, the organizing activity of living systems, at all levels of life, is mental activity. The interactions of a living organism—plant, animal, or human—with its environment are cognitive, or mental interactions. Thus life and cognition become inseparably connected. Mind—or, more accurately, mental process—is immanent in matter at all levels of life.

The new concept of mind was developed independently by Gregory Bateson and Humberto Maturana during the 1960s. Bateson, who was a regular participant in the legendary Macy Conferences during the early years of cybernetics, pioneered the application of systems thinking and cybernetic principles in several areas.[15] In particular he developed a systems approach to mental illness and a cybernetic model of alcoholism, which led him to define "mental process" as a systems phenomenon characteristic of living organisms.

Bateson listed a set of criteria that systems have to satisfy for mind to occur.[16] Any system that satisfies those criteria will be able to develop the processes we associate with mind—learning, memory, decision making, and so on. In Bateson's view these mental processes are a necessary and inevitable consequence of a certain complexity that begins long before organisms develop brains and higher nervous systems. He also emphasized that mind is manifest not only in individual organisms, but also in social systems and ecosystems.

Bateson presented his new concept of mental process for the first time in 1969 in Hawaii, in a paper he gave at a conference on mental health.[17] This was the very year in which Maturana presented a different formulation of the same basic idea at the conference on cognition organized by Heinz von Foerster in Chicago.[18] Thus two scientists, both strongly influenced by cybernetics, had arrived simultaneously at the same revolutionary concept of mind. However, their methods were quite different, as were the languages in which they described their groundbreaking discovery.

Bateson's whole thinking was in terms of patterns and relationships. His main aim, like Maturana's, was to discover the pattern of organization common to all living creatures. "What pattern," he asked, "connects the crab to the lobster and the orchid to the primrose and all four of them to me? And me to you?"[19]

Bateson thought that in order to describe nature accurately one should try to speak nature's language, which, he insisted, is a language of relationships. Relationships are the essence of the living world, according to Bateson. Biological form consists of relationships, not of parts, and he emphasized that this is also how people think. Therefore he called the book in which he discussed his concept of mental process *Mind and Nature: A Necessary Unity.*

Bateson had a unique ability to glean insights from nature by intense observation. This was not just ordinary scientific observation. He was able, somehow, to observe a plant or animal with his whole being, with empathy and passion. And when he talked about it he would describe that plant in minute and loving detail, using what he considered to be the language of nature to talk about the general principles he derived from his direct contact with the plant. He was very taken by the beauty manifest in the complexity of nature's patterned relationships, and the description of these patterns gave him a strong aesthetic pleasure.

Bateson developed his criteria of mental process intuitively from his keen observation of the living world. It was clear to him that the phenomenon of mind was inseparably connected with the phenomenon of life. When he looked at the living world, he saw its organizing activity as being essentially mental. In his own words, "mind is the essence of being alive."[20]

In spite of his clear recognition of the unity of mind and life—or mind and nature, as he would put it—Bateson never asked, What is life? He never felt the need to develop a theory, or even a model, of living systems that would provide a conceptual framework for his criteria of mental process. To develop such a framework was precisely Maturana's approach.

By coincidence—or perhaps intuition?—Maturana struggled simultaneously with two questions that seemed to him to lead in opposite directions: What is the nature of life? and What is cognition?[21] Eventually he discovered that the answer to the first question—autopoiesis—provided him with the theoretical framework for answering the second. The result is a systems theory of cognition, developed by Maturana and Varela, which is sometimes called the Santiago theory.

The central insight of the Santiago theory is the same as Bateson's—the identification of cognition, the process of knowing, with the process of life.[22] This represents a radical expansion of the traditional concept of mind. According to the Santiago theory, the brain is not necessary for mind to exist. A bacterium, or a plant, has no brain but has a mind. The simplest organisms are capable of perception and thus of cognition. They do not see, but they nevertheless perceive changes in their environment—differences between light and shadow, hot and cold, higher and lower concentrations of some chemical, and the like.

The new concept of cognition, the process of knowing, is thus much broader than that of thinking. It involves perception, emotion, and action—the entire process of life. In the human realm cognition also includes language, conceptual thinking, and all the other attributes of human consciousness. The general concept, however, is much broader and does not necessarily involve thinking.

The Santiago theory provides, in my view, the first coherent scientific framework that really overcomes the Cartesian split. Mind and matter no longer appear to belong to two separate categories but are seen as representing merely different aspects, or dimensions, of the same phenomenon of life.

To illustrate the conceptual advance represented by this unified view of mind, matter, and life, let us turn to a question that has confused scientists and philosophers for over a hundred years: What is the relationship between the mind and the brain? Neuroscientists have known since the nineteenth century that brain structures and mental functions are intimately connected, but the exact relationship between mind and brain always remained a mystery. As recently as 1994 the editors of an anthology titled *Consciousness in Philosophy and Cognitive Neuroscience* stated frankly in their introduction: "Even though everybody agrees that mind has something to do with the brain, there is still no general agreement on the exact nature of this relationship."[23]

In the Santiago theory the relationship between mind and brain is simple and clear. Descartes's characterization of mind as "the thinking thing" *(res cogitans)* is finally abandoned. Mind is not a thing but a process—the process of cognition, which is identified with the process of life. The brain is a specific structure through which this process operates. The relationship between mind and brain, therefore, is one between process and structure.

The brain is, of course, not the only structure through which the process of cognition operates. The entire dissipative structure of the organism participates in the process of cognition, whether or not the organism has a brain and a higher nervous system. Moreover, recent research indicates strongly that in the human organism the nervous system, the immune system, and the endocrine system, which traditionally have been viewed as three separate systems, in fact form a single cognitive network.[24]

The new synthesis of mind, matter, and life, which will be explored in great detail in the following pages, involves two conceptual unifications. The interdependence of pattern and structure allows us to integrate two approaches to the understanding of nature that have been separate and in competition throughout Western science and philosophy. The interdependence of process and structure allows us to heal the split between mind and matter that has haunted our modern era ever since Descartes. Together these two unifications provide the three interdependent conceptual dimensions for the new scientific understanding of life.

8

Dissipative Structures

Structure and Change

Since the early days of biology, philosophers and scientists have noticed that living forms, in many seemingly mysterious ways, combine the stability of structure with the fluidity of change. Like whirlpools, they depend on a constant flow of matter through them; like flames, they transform the materials on which they feed to maintain their activities and to grow; but unlike whirlpools or flames, living structures also develop, reproduce, and evolve.

In the 1940s Ludwig von Bertalanffy called such living structures "open systems" to emphasize their dependence on continual flows of energy and resources. He coined the term *Fliessgleichgewicht* ("flowing balance") to express the coexistence of balance and flow, of structure and change, in all forms of life.[1] Subsequently ecologists began to picture ecosystems in terms of flow diagrams, mapping out the pathways of energy and matter in various food webs. These studies established recycling as a key principle of ecology. Being open systems, all organisms in an ecosystem produce wastes, but what is waste for one species is food for another, so that wastes are continually recycled and the ecosystem as a whole generally remains without waste.

Green plants play a vital role in the flow of energy through all ecological cycles. Their roots take in water and mineral salts from the earth, and the resulting juices rise up to the leaves, where they combine with carbon dioxide (CO_2) from the air to form sugars and other organic compounds. (These include cellulose, the main structural element of cell walls.) In this marvelous process, known as photosynthesis, solar energy is converted into chemical energy and bound in the organic substances, while oxygen is released into the air to be taken up again by other plants, and by animals, in the process of respiration.

By blending water and minerals from below with sunlight and CO_2 from above, green plants link the earth and the sky. We tend to believe that plants grow out of the soil, but in fact most of their substance comes from the air. The bulk of the cellulose and the other organic compounds produced through photosynthesis consists of heavy carbon and oxygen atoms, which plants take directly from the air in the form of CO_2. Thus the weight of a wooden log comes almost entirely from the air. When we burn a log in a fireplace, oxygen and carbon combine once more into CO_2, and in the light and heat of the fire we recover part of the solar energy that went into making the wood.

Figure 8-1 shows a picture of a typical food cycle. As plants are eaten by animals, which in turn are eaten by other animals, the plants' nutrients are passed on through the food web, while energy is dissipated as heat through respiration and as waste through excretion. The wastes, as well as dead animals and plants, are decomposed by so-called decomposer organisms (insects and bacteria), which break them down into basic nutrients, to be taken up once more by green plants. In this way nutrients and other basic elements continually cycle through the ecosystem, while energy is dissipated at each stage. Thus Eugene Odum's dictum "Matter circulates, energy dissipates."[2] The only waste generated by the ecosystem as a whole is the heat energy of respiration, which is radiated into the atmosphere and is replenished continually by the sun through photosynthesis.

Our illustration is, of course, greatly simplified. The actual food cycles can be understood only within the context of much more

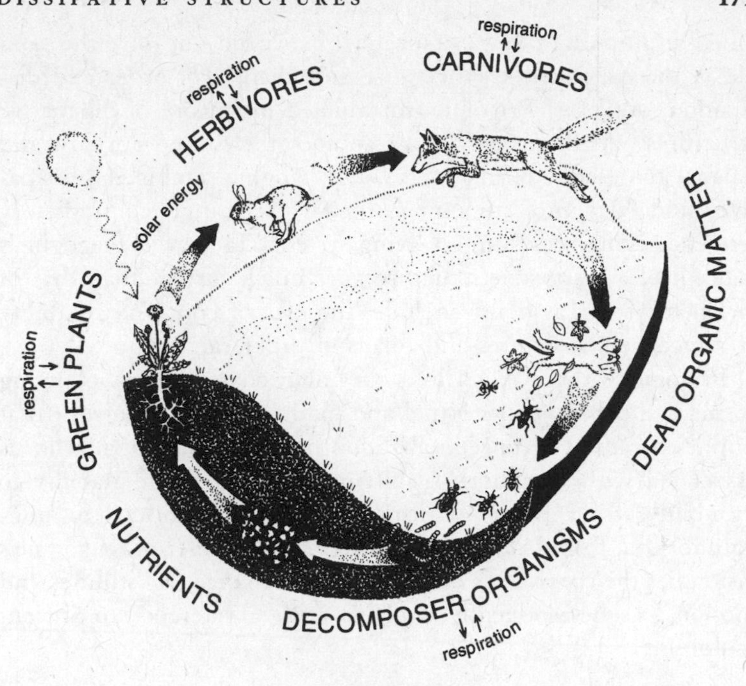

Figure 8-1
A typical food cycle.

complex food webs in which the basic nutrient elements appear in a variety of chemical compounds. In recent years our knowledge of those food webs has been expanded and refined considerably by the Gaia theory, which shows the complex interweaving of living and nonliving systems throughout the biosphere—plants and rocks, animals and atmospheric gases, microorganisms and oceans.

The flow of nutrients through an ecosystem's organisms, moreover, is not always smooth and even, but often proceeds in pulses, jolts, and floods. In the words of Prigogine and Stengers, "The energy flow that crosses [an organism] somewhat resembles the flow of a river that generally moves smoothly but from time to time tumbles down a waterfall, which liberates part of the energy it contains."[3]

The understanding of living structures as open systems provided an important new perspective, but it did not solve the puzzle of the coexistence of structure and change, of order and dissipation, until Ilya Prigogine formulated his theory of dissipative structures.[4] As Bertalanffy had combined the concepts of flow and balance to describe open systems, so Prigogine combined "dissipative" and "structure" to express the two seemingly contradictory tendencies that coexist in all living systems. However, Prigogine's concept of a dissipative structure goes much further than that of an open system, as it also includes the idea of points of instability at which new structures and forms of order can emerge.

Prigogine's theory interlinks the main characteristics of living forms in a coherent conceptual and mathematical framework that implies a radical reconceptualization of many fundamental ideas associated with structure—a shift of perception from stability to instability, from order to disorder, from equilibrium to nonequilibrium, from being to becoming. At the center of Prigogine's vision lies the coexistence of structure and change, of "stillness and motion," as he eloquently explains with a reference to ancient sculpture:

> Each great period of science has led to some model of nature. For classical science it was the clock; for nineteenth-century science, the period of the Industrial Revolution, it was an engine running down. What will be the symbol for us? What we have in mind may perhaps be expressed by a reference to sculpture, from Indian or pre-Columbian art to our time. In some of the most beautiful manifestations of sculpture, be it the dancing Shiva or in the miniature temples of Guerrero, there appears very clearly the search for a junction between stillness and motion, time arrested and time passing. We believe that this confrontation will give our period its uniqueness.[5]

Nonequilibrium and Nonlinearity

The key to understanding dissipative structures is to realize that they maintain themselves in a stable state far from equilibrium. This situation is so different from the phenomena described by classical science that we run into difficulties with conventional language. Dictionary definitions of the word "stable" include "fixed," "not fluctuating," and "unvarying," all of which are inaccurate to describe dissipative structures. A living organism is characterized by continual flow and change in its metabolism, involving thousands of chemical reactions. Chemical and thermal equilibrium exists when all these processes come to a halt. In other words, an organism in equilibrium is a dead organism. Living organisms continually maintain themselves in a state far from equilibrium, which is the state of life. Although very different from equilibrium, this state is nevertheless stable over long periods of time, which means that, as in a whirlpool, the same overall structure is maintained in spite of the ongoing flow and change of components.

Prigogine realized that classical thermodynamics, the first science of complexity, is inappropriate to describe systems far from equilibrium because of the linear nature of its mathematical structure. Close to equilibrium—in the range of classical thermodynamics—there are flow processes, called "fluxes," but they are weak. The system will always evolve toward a stationary state in which the generation of entropy (or disorder) is as small as possible. In other words, the system will minimize its fluxes, staying as close as possible to the equilibrium state. In this range the flow processes can be described by linear equations.

Farther away from equilibrium, the fluxes are stronger, entropy production increases, and the system no longer tends toward equilibrium. On the contrary, it may encounter instabilities leading to new forms of order that move the system farther and farther away from the equilibrium state. In other words, far from equilibrium, dissipative structures may develop into forms of ever-increasing complexity.

Prigogine emphasizes that the characteristics of a dissipative structure cannot be derived from the properties of its parts but are consequences of "supramolecular organization."[6] Long-range correlations appear at the precise point of transition from equilibrium to nonequilibrium, and from that point on the system behaves as a whole.

Far from equilibrium, the system's flow processes are interlinked through multiple feedback loops, and the corresponding mathematical equations are nonlinear. The farther a dissipative structure is from equilibrium, the greater is its complexity and the higher is the degree of nonlinearity in the mathematical equations describing it.

Recognizing the crucial link between nonequilibrium and nonlinearity, Prigogine and his collaborators developed a nonlinear thermodynamics for systems far from equilibrium, using the techniques of dynamical systems theory, the new mathematics of complexity, which was just being developed.[7] The linear equations of classical thermodynamics, Prigogine noted, can be analyzed in terms of point attractors. Whatever the system's initial conditions, it will be "attracted" toward a stationary state of minimum entropy, as close to equilibrium as possible, and its behavior will be completely predictable. As Prigogine puts it, systems in the linear range tend to "forget their initial conditions."[8]

Outside the linear region the situation is dramatically different. Nonlinear equations usually have more than one solution; the higher the nonlinearity, the greater the number of solutions. This means that new situations may emerge at any moment. Mathematically speaking, the system encounters a bifurcation point in such a case, at which it may branch off into an entirely new state. We shall see below that the behavior of the system at the bifurcation point (in other words, which one of several available new branches it will take) depends on the previous history of the system. In the nonlinear range initial conditions are no longer "forgotten."

Moreover, Prigogine's theory shows that the behavior of a dissipative structure far from equilibrium no longer follows any universal law but is unique to the system. Near equilibrium we find repetitive phenomena and universal laws. As we move away from equilibrium, we move from the universal to the unique, toward richness and variety. This, of course, is a well-known characteristic of life.

The existence of bifurcations at which the system may take several different paths implies that indeterminacy is another characteristic of Prigogine's theory. At the bifurcation point the system can "choose"—the term is used metaphorically—from among several possible paths, or states. Which path it will take will depend on the system's history and on various external conditions and can never be predicted. There is an irreducible random element at each bifurcation point.

This indeterminacy at bifurcation points is one of two kinds of unpredictability in the theory of dissipative structures. The other kind, which is also present in chaos theory, is due to the highly nonlinear nature of the equations and exists even when there are no bifurcations. Because of repeated feedback loops—or, mathematically, repeated iterations—the tiniest error in the calculations, caused by the practical need to round off figures at some decimal point, will inevitably add up to sufficient uncertainty to make predictions impossible.[9]

The indeterminacy at the bifurcation points and the "chaos-type" unpredictability due to repeated iterations both imply that the behavior of a dissipative structure can be predicted only over a short time span. After that, the system's trajectory eludes us. Thus Prigogine's theory, like quantum theory and chaos theory, reminds us once more that scientific knowledge offers but "a limited window on the universe."[10]

The Arrow of Time

According to Prigogine, the recognition of indeterminacy as a key characteristic of natural phenomena is part of a profound reconceptualization of science. A closely related aspect of this conceptual shift concerns the scientific notions of irreversibility and time.

In the mechanistic paradigm of Newtonian science, the world was seen as completely causal and determinate. All that happened had a definite cause and gave rise to a definite effect. The future of any part of the system, as well as its past, could in principle be calculated with absolute certainty if its state at any given time was known in all details. This rigorous determinism found its clearest expression in the celebrated words of Pierre Simon Laplace:

> An intellect which at a given instant knew all the forces acting in nature, and the position of all things of which the world consists—supposing the said intellect were vast enough to subject these data to analysis—would embrace in the same formula the motions of the greatest bodies in the universe and those of the slightest atoms; nothing would be uncertain for it, and the future, like the past, would be present to its eyes.[11]

In this Laplacian determinism, there is no difference between the past and the future. Both are implicit in the present state of the world and in the Newtonian equations of motion. All processes are strictly reversible. Both future and past are interchangeable; there is no room for history, novelty, or creativity.

Irreversible effects (such as friction) were noticed in classical Newtonian physics, but they were always neglected. In the nineteenth century this situation changed dramatically. With the invention of thermal engines, the irreversibility of energy dissipation in friction, viscosity (the resistance of a fluid to flow), and heat losses became the central focus of the new science of thermodynamics, which introduced the idea of an "arrow of time." Concurrently, geologists, biologists, philosophers, and poets all began to think about change, growth, development, and evolution. Nineteenth-century thought was deeply concerned with the nature of becoming.

In classical thermodynamics irreversibility, although an important feature, is always associated with energy losses and waste. Prigogine introduced a fundamental change of this view in his theory of dissipative structures by showing that in living systems, which operate far from equilibrium, irreversible processes play a constructive and indispensable role.

Chemical reactions, the basic processes of life, are the prototype of irreversible processes. In a Newtonian world there would be no chemistry and no life. Prigogine's theory shows how a particular type of chemical processes, the catalytic loops that are essential to living organisms,[12] lead to instabilities through repeated self-amplifying feedback, and how new structures of ever-increasing complexity emerge at successive bifurcation points. "Irreversibility,"

Prigogine concluded, "is the mechanism that brings order out of chaos."[13]

Thus the conceptual shift in science advocated by Prigogine is one from deterministic reversible processes to indeterminate and irreversible ones. Since the irreversible processes are essential to chemistry and to life, while the interchangeability of the future and the past is an integral part of physics, it seems that Prigogine's reconceptualization must be seen in the larger context discussed at the beginning of this book in connection with deep ecology, as part of the paradigm shift from physics to the life sciences.[14]

Order and Disorder

The arrow of time introduced in classical thermodynamics did not point toward increasing order; it pointed away from it. According to the second law of thermodynamics, there is a trend in physical phenomena from order to disorder, toward ever-increasing entropy.[15] One of Prigogine's greatest achievements has been to resolve the paradox of the two contradictory views of evolution in physics and biology—one of an engine running down, the other of a living world unfolding toward increasing order and complexity. In Prigogine's own words, "There is [a] question, which has plagued us for more than a century: What significance does the evolution of a living being have in the world described by thermodynamics, a world of ever-increasing disorder?"[16]

In Prigogine's theory the second law of thermodynamics is still valid, but the relationship between entropy and disorder is seen in a new light. To understand this new perception it is helpful to review the classical definitions of entropy and order. The concept of entropy was introduced in the nineteenth century by Rudolf Clausius, a German physicist and mathematician, to measure the dissipation of energy into heat and friction. Clausius defined the entropy generated in a thermal process as the dissipated energy divided by the temperature at which the process takes place. According to the second law, that entropy keeps increasing as the thermal process continues; the dissipated energy can never be recovered; and this direction toward ever-increasing entropy defines the arrow of time.

Although the dissipation of energy into heat and friction is a common experience, a puzzling question arose as soon as the second law was formulated: What exactly causes this irreversibility? In Newtonian physics the effects of friction had usually been neglected because they were not considered very important. However, these effects *can* be taken into account within the Newtonian framework. In principle, scientists argued, one should be able to use Newton's laws of motion to describe the dissipation of energy at the level of molecules in terms of cascades of collisions. Each of these collisions is a reversible event, so it should be perfectly possible to run the whole process backward. The dissipation of energy, which is irreversible at the macroscopic level, according to the second law and to common experience, seems to be composed of completely reversible events at the microscopic level. So where does irreversibility creep in?

This mystery was solved at the turn of the century by the Austrian physicist Ludwig Boltzmann, one of the great theorists of classical thermodynamics, who gave a new meaning to the concept of entropy and established the link between entropy and order. Following a line of reasoning developed originally by James Clerk Maxwell, the founder of statistical mechanics,[17] Boltzmann devised an ingenious thought experiment to examine the concept of entropy at the molecular level.[18]

Suppose we have a box, Boltzmann reasoned, divided into two equal compartments by an imaginary partition at the center, and eight distinguishable molecules, numbered from one to eight like billiard balls. How many ways are there to distribute these particles in the box in such a way that a certain number of them are on the left side of the partition and the rest on the right?

First, let us put all eight particles on the left side. There is only one way of doing that. However, if we put seven particles on the left and one on the right, there are eight different possibilities, because the single particle on the right side of the box may be each of the eight particles in turn. Since the molecules are distinguishable, these eight possibilities all count as different arrangements.

Similarly, there are twenty-eight different arrangements for six particles on the left and two on the right.

A general formula for all these permutations can easily be de-

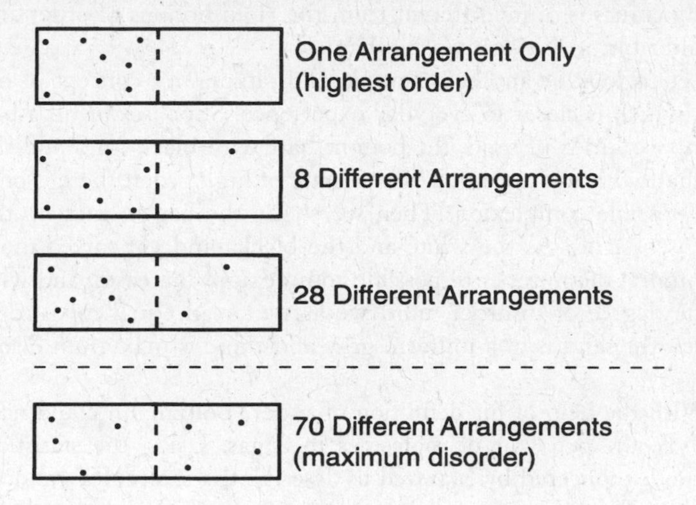

Figure 8-2
Boltzmann's thought experiment.

rived.[19] It shows that the number of possibilities increases as the difference between the numbers of particles on the left and right becomes smaller, reaching a maximum of seventy different arrangements when there is an equal distribution of molecules, four on each side (see figure 8-2).

Boltzmann called the different arrangements "complexions" and associated them with the concept of order—the lower the number of complexions, the higher the order. Thus, in our example, the first state with all eight particles on one side displays the highest order, while the equal distribution with four particles on each side represents the maximum disorder.

It is important to emphasize that the concept of order introduced by Boltzmann is a thermo-*dynamic* concept, where the molecules are in constant motion. In our example the partition of the box is purely imaginary, and molecules in random motion will keep going across it. Over time the gas will be in different states—

that is, with different numbers of molecules on the two sides of the box—and the number of complexions for each of these states is related to its degree of order. This definition of order in thermodynamics is quite different from the rigid notions of order and equilibrium in Newtonian mechanics.

Let us look at another example of Boltzmann's concept of order, which is closer to everyday experience. Suppose we fill a bag with two kinds of sand, the bottom half with black sand and the top half with white sand. This is a state of high order; there is only one possible complexion. Then we shake the bag to mix up the grains of sand. As the white and the black sand get mixed more and more, the number of possible complexions increases, and with it the degree of disorder, until we arrive at an equal mixture in which the sand is of a uniform gray and there is maximum disorder.

With the help of his definition of order, Boltzmann could now analyze the behavior of molecules in a gas. Using the statistical methods pioneered by Maxwell to describe the molecules' random motion, Boltzmann noted that the number of possible complexions of any state measures the probability of the gas being in that state. This is how probability is defined. The more complexions there are for a certain arrangement, the more likely will that state occur in a gas with molecules in random motion.

Thus the number of possible complexions for a certain arrangement of molecules measures both the degree of order of that state and the probability of its occurrence. The higher the number of complexions, the greater will the disorder be, and the more likely the gas will be in that state. Boltzmann therefore concluded that the movement from order to disorder is a movement from an unlikely state to a likely state. By identifying entropy and disorder with the number of complexions, he introduced a definition of entropy in terms of probabilities.

According to Boltzmann, there is no law of physics that forbids a movement from disorder to order, but with a random motion of molecules such a direction is very unlikely. The larger the number of molecules, the higher the probability of movement from order to disorder, and with the enormous number of particles in a gas that probability, for all practical purposes, becomes certainty. When you shake a bag with white and black sand, you may observe the two kinds of grains drift apart, seemingly miraculously, to create the highly ordered state of complete separation. But you are likely to have to shake the bag for a few million years for that event to happen.

In Boltzmann's language the second law of thermodynamics means that any closed system will tend toward the state of maximum probability, which is a state of maximum disorder. Mathematically this state can be defined as the attractor state of thermal equilibrium. Once equilibrium has been reached, the system is not likely to move away from it. At times the molecules' random motion will result in different states, but these will be close to equilibrium and will exist only for short periods of time. In other words, the system will merely fluctuate around the state of thermal equilibrium.

Classical thermodynamics, then, is appropriate to describe phenomena at equilibrium or close to equilibrium. Prigogine's theory of dissipative structures, by contrast, applies to thermodynamic phenomena far from equilibrium, where molecules are not in random motion but are interlinked through multiple feedback loops, described by nonlinear equations. These equations are no longer dominated by point attractors, which means that the system no longer tends toward equilibrium. A dissipative structure maintains itself far from equilibrium and may even move farther and farther away from it through a series of bifurcations.

At the bifurcation points, states of higher order (in Boltzmann's sense) may emerge spontaneously. However, this does not contradict the second law of thermodynamics. The total entropy of the system keeps increasing, but this increase in entropy is not a uniform increase in disorder. In the living world order and disorder are always created simultaneously.

According to Prigogine, dissipative structures are islands of order in a sea of disorder, maintaining and even increasing their order at the expense of greater disorder in their environment. For example, living organisms take in ordered structures (food) from their environment, use them as resources for their metabolism, and dissipate structures of lower order (waste). In this way order "floats in disorder," as Prigogine puts it, while the overall entropy keeps increasing in accordance with the second law.[20]

This new perception of order and disorder represents an inversion of traditional scientific views. According to the classical view, for which physics was the principal source of concepts and metaphors, order is associated with equilibrium, as, for example, in crystals and other static structures, and disorder with nonequilibrium situations, such as turbulence. In the new science of complexity, which takes its inspiration from the web of life, we learn that nonequilibrium is a source of order. The turbulent flows of water and air, while appearing chaotic, are really highly organized, exhibiting complex patterns of vortices dividing and subdividing again and again at smaller and smaller scales. In living systems the order arising from nonequilibrium is far more evident, being manifest in the richness, diversity, and beauty of life all around us. Throughout the living world chaos is transformed into order.

Points of Instability

The points of instability at which dramatic and unpredictable events take place, where order emerges spontaneously and complexity unfolds, are perhaps the most intriguing and fascinating aspect of the theory of dissipative structures. Before Prigogine, the only type of instability studied in some detail was that of turbulence, caused by the internal friction of a flowing liquid or gas.[21] Leonardo da Vinci made many careful studies of turbulent flows of water, and in the nineteenth century a series of experiments was undertaken that showed that any flow of water or air will become turbulent at sufficiently high velocity—in other words, at sufficiently large "distance" from equilibrium (the motionless state).

Prigogine's studies showed that this is not true for chemical reactions. Chemical instabilities will not automatically appear far from equilibrium. They require the presence of catalytic loops, which bring the system to the point of instability through repeated self-amplifying feedback.[22] These processes combine two different phenomena: chemical reactions and diffusion (the physical flow of molecules due to differences in concentration). Accordingly, the nonlinear equations describing them are called "reaction-diffusion equations." They form the mathematical core of Prigogine's theory, allowing for an astonishing range of behaviors.[23]

The British biologist Brian Goodwin has applied Prigogine's mathematical techniques in a most ingenious way to model the stages of development of a very special single-celled alga.[24] By setting up differential equations that interrelate patterns of calcium concentration in the alga's cell fluid with the mechanical properties of the cell walls, Goodwin and his colleagues were able to identify feedback loops in a self-organizing process, in which structures of increasing order emerge at successive bifurcation points.

A bifurcation point is a threshold of stability at which the dissipative structure may either break down or break through to one of several new states of order. What exactly happens at this critical point depends on the system's previous history. Depending on which path it has taken to reach the point of instability, it will follow one or another of the available branches after the bifurcation.

This important role of the history of a dissipative structure at critical points of its further development, which Prigogine has observed even in simple chemical oscillations, seems to be the physical origin of the connection between structure and history that is characteristic of all living systems. Living structure, as we shall see, is always a record of previous development.[25]

At the bifurcation point, the dissipative structure also shows an extraordinary sensitivity to small fluctuations in its environment. A tiny random fluctuation, often called "noise," can induce the choice of path. Since all living systems exist in continually fluctuating environments, and since we can never know which fluctuation will occur at the bifurcation point just at the "right" moment, we can never predict the future path of the system.

Thus all deterministic description breaks down when a dissipative structure crosses the bifurcation point. Minute fluctuations in the environment will lead to the choice of the branch it will follow. And since, in a sense, it is those random fluctuations that lead to the emergence of new forms of order, Prigogine has coined the phrase "order through fluctuations" to describe the situation.

The equations of Prigogine's theory are deterministic equations. They govern the system's behavior between bifurcation points, while random fluctuations are decisive at the points of instability. Thus "self-organization processes in far-from-equilibrium conditions correspond to a delicate interplay between chance and necessity, between fluctuations and deterministic laws."[26]

A New Dialogue with Nature

The conceptual shift implied in Prigogine's theory involves several closely interrelated ideas. The description of *dissipative structures* that exist *far from equilibrium* requires a *nonlinear* mathematical formalism, capable of modeling multiple interlinked feedback loops. In living organisms these are catalytic loops (that is, nonlinear, *irreversible* chemical processes), which lead to *instabilities* through repeated self-amplifying feedback. When a dissipative structure reaches such a point of instability, called a *bifurcation point,* an element of *indeterminacy* enters into the theory. At the bifurcation point the system's behavior is inherently *unpredictable.* In particular, new structures of higher *order* and complexity may emerge spontaneously. Thus self-organization, the spontaneous emergence of order, results from the combined effects of nonequilibrium, irreversibility, feedback loops, and instability.

The radical nature of Prigogine's vision is apparent from the fact that these fundamental ideas were rarely addressed in traditional science and were often given negative connotations. This is evident in the very language used to express them. *Non*-equilibrium, *non*linearity, *in*stability, *in*determinacy, and so on, are all negative formulations. Prigogine believes that the conceptual shift implied by his theory of dissipative structures is not only crucial for scientists to understand the nature of life, but will also help us to integrate ourselves more fully into nature.

Many of the key characteristics of dissipative structures—the sensitivity to small changes in the environment, the relevance of previous history at critical points of choice, the uncertainty and unpredictability of the future—are revolutionary new concepts from the point of view of classical science but are an integral part of human experience. Since dissipative structures are the basic structures of all living systems, including human beings, this should perhaps not come as a great surprise.

Instead of being a machine, nature at large turns out to be more like human nature—unpredictable, sensitive to the surrounding world, influenced by small fluctuations. Accordingly, the appropriate way of approaching nature to learn about her complexity and beauty is not through domination and control, but through respect, cooperation, and dialogue. Indeed, Ilya Prigogine and Isabelle Stengers gave their popular book, *Order out of Chaos,* the subtitle "Man's New Dialogue with Nature."

In the deterministic world of Newton there is no history and no creativity. In the living world of dissipative structures history plays an important role, the future is uncertain, and this uncertainty is at the heart of creativity. "Today," Prigogine reflects, "the world we see outside and the world we see within are converging. This convergence of two worlds is perhaps one of the important cultural events of our age."[27]

9

Self-Making

Cellular Automata

When Ilya Prigogine developed his theory of dissipative structures, he looked for the simplest examples he could describe mathematically. He found them in the catalytic loops of chemical oscillations, also known as "chemical clocks."[1] These are not living systems, but the same kinds of catalytic loops are central to the metabolism of a cell, the simplest known living system. Therefore Prigogine's model allows us to understand the essential structural features of cells in terms of dissipative structures.

Humberto Maturana and Francisco Varela followed a similar strategy when they developed their theory of autopoiesis, the pattern of organization of living systems.[2] They asked themselves: What is the simplest embodiment of an autopoietic network that can be described mathematically? Like Prigogine, they found that even the simplest cell was too complex for a mathematical model. On the other hand, they also realized that since the pattern of autopoiesis is the defining characteristic of a living system, there is no autopoietic system in nature simpler than a cell. So instead of looking for a natural autopoietic system, they decided to simulate one with a computer program.

Their approach was analogous to the Daisyworld model James Lovelock designed several years later.[3] But where Lovelock looked for the simplest mathematical simulation of a planet with a biosphere that would regulate its temperature, Maturana and Varela looked for the simplest simulation of a network of cellular processes embodying an autopoietic pattern of organization. This meant that they had to design a computer program simulating a network of processes, in which the function of each component is to help produce or transform other components in the network. As in a cell, this autopoietic network would also have to create its own boundary, which would participate in the network of processes and at the same time define its extension.

To find an appropriate mathematical technique for this task, Francisco Varela examined the mathematical models of self-organizing networks developed in cybernetics. The binary networks pioneered by McCulloch and Pitts in the 1940s did not offer sufficient complexity to simulate an autopoietic network,[4] but subsequent network models, known as "cellular automata," turned out to provide the ideal techniques.

A cellular automaton is a rectangular grid of regular squares, or "cells," like a chess board. Each cell can take on a number of different values and has a definite number of neighbor cells that can influence it. The pattern, or "state," of the entire grid changes in discrete steps according to a set of "transition rules" that apply simultaneously to every cell. Cellular automata are usually assumed to be completely deterministic, but random elements can easily be introduced into the rules, as we shall see.

These mathematical models are called "automata" because they were invented originally by John von Neumann to construct self-duplicating machines. Although such machines were never built, von Neumann showed in an abstract and elegant way that, in principle, this could be done.[5] Since then, cellular automata have been widely used both to model natural systems and to invent a large number of mathematical games.[6] Perhaps the best-known example is the game "Life," in which each cell can have one of two values—say, "black" or "white"—and the sequence of states is determined by three simple rules, called "birth," "death," and "survival."[7] The game can produce an amazing variety of patterns. Some of them "move"; others remain stable; yet other patterns oscillate or behave in more complex manners.[8]

While cellular automata were used by professional and amateur mathematicians to invent numerous games, they were also studied extensively as mathematical tools for scientific models. Because of their network structure and their ability to accommodate large numbers of discrete variables, these mathematical forms were soon recognized as an exciting alternative to differential equations for modeling complex systems.[9] In a sense, the two approaches—differential equations and cellular automata—can be seen as different mathematical frameworks corresponding to the two distinct conceptual dimensions—structure and pattern—of the theory of living systems.

Simulating Autopoietic Networks

In the early 1970s Francisco Varela realized that the step-by-step sequences of cellular automata, which are ideal for computer simulations, provided him with a powerful tool for simulating autopoietic networks. Indeed, in 1974 Varela succeeded in constructing the appropriate computer simulation together with Maturana and computer scientist Ricardo Uribe.[10] Their cellular automaton consists of a grid in which a "catalyst" and two kinds of elements move randomly and interact with one another in such a way that further elements of both kinds may be produced; others may disappear, and certain elements may bond with each other to form chains.

In the computer printouts of the grid, the "catalyst" is marked by a star (⋆). The first kind of element, which is present in great numbers, is called a "substrate element" and is marked by a circle (o); the second kind is called a "link" and is marked by a circle inside a square (⊡). There are three different kinds of interactions and transformations. Two substrate elements may coalesce in the presence of the catalyst to produce a link; several links may "bond"—that is, they may stick together—to form a chain; and any link, either free or bonded in a chain, may disintegrate again into two substrate elements. Eventually a chain may also close upon itself.

The three interactions are defined symbolically as follows.

1. Production: * + O + O —> * + [O]

2. Bonding: [O] + [O] —> [O]-[O]

[O]-[O] + [O] —> [O]-[O]-[O]

etc.

3. Disintegration: [O] —> O + O

The exact mathematical prescriptions (the so-called algorithm) for when and how these processes take place are quite elaborate. They consist of numerous rules for the movements of the various elements and for their mutual interactions.[11] For example, the rules for motion include the following:

- Substrate elements are allowed to move only into unoccupied spaces ("holes") in the grid, while the catalyst and the links may displace substrate elements, pushing them into adjacent holes. The catalyst may similarly displace a free link.
- The catalyst and the links may also exchange places with a substrate element and thus can pass freely through the substrate.
- Substrate elements, but not the catalyst or the free links, may pass through a chain to occupy a hole behind it. (This simulates the semipermeable membranes of cells.)
- Bonded links in a chain cannot move at all.

Within these rules the actual motion of the elements and many details of their mutual interactions—production, bonding, and disintegration—are chosen at random.[12] When the simulation is run on a computer, a network of interactions is generated, which involves many random choices and thus may generate many different sequences. The authors were able to show that some of those sequences generate stable autopoietic patterns.

An example of such a sequence from their paper, shown in

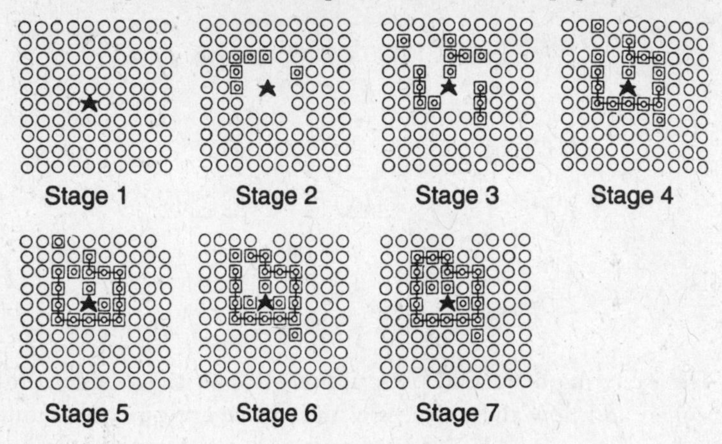

Figure 9-1
Computer simulation of autopoietic network.

seven stages, is reproduced in figure 9-1. In the initial state (stage one) one space in the grid is occupied by the catalyst and all the others by the substrate elements. In stage two several links have been produced and, accordingly, there are now several holes in the grid. In stage three more links have been produced and some of them have bonded. The production of links and the formation of bonds both increase as the simulation proceeds through stages four to six, and in stage seven we see that the chain of bonded links has closed upon itself, enclosing the catalyst, three links, and two substrate elements. Thus the chain has formed an enclosure that is penetrable for the substrate elements but not for the catalyst. Whenever such a situation occurs, the closed chain may stabilize itself and become the boundary of an autopoietic network. Indeed, this happened in this particular sequence. Subsequent stages of the computer run showed that occasionally some links in the boundary would disintegrate, but that these would eventually be replaced by new links produced inside the enclosure in the presence of the catalyst.

In the long run the chain continued to form an enclosure for the catalyst, while its links kept disintegrating and being replaced. In this way the membrane-like chain became the boundary of a network of transformations while at the same time participating in that network of processes. In other words, an autopoietic network was simulated.

Whether or not a sequence of this simulation will generate an autopoietic pattern depends crucially on the disintegration probability—that is, on how often links will disintegrate. Since the delicate balance of disintegration and "repair" is based on random motion of substrate elements through the membrane, random production of new links, and random motion of those new links to the repair site, the membrane will remain stable only if all those processes are likely to be completed before further disintegrations occur. The authors showed that with very small disintegration probabilities viable autopoietic patterns can indeed be achieved.[13]

Binary Networks

The cellular automaton designed by Varela and his colleagues was one of the first examples of how the self-organizing networks of living systems can be simulated. Over the past twenty years many other simulations have been studied, and it has been demonstrated that these mathematical models can spontaneously generate complex and highly ordered patterns, exhibiting some important principles of the order found in living systems.

These studies intensified when it was recognized that the newly developed techniques of dynamical systems theory—attractors, phase portraits, bifurcation diagrams, and so on—can be used as effective tools to analyze the mathematical network models. Equipped with these new techniques, scientists once more studied the binary networks developed in the 1940s and found that even though these are not autopoietic networks, their analysis leads to surprising insights about the network patterns of living systems. Much of this work has been carried out by evolutionary biologist Stuart Kauffman and his colleagues at the Santa Fe Institute in New Mexico.[14]

Since the study of complex systems with the help of attractors and phase portraits is very much associated with the development of chaos theory, it was natural for Kauffman and his colleagues to ask: What is the role of chaos in living systems? We are still far from a full answer to this question, but Kauffman's work has resulted in some very exciting ideas. To understand these, we need to take a closer look at binary networks.

A binary network consists of nodes capable of two distinct values, conventionally labeled ON and OFF. It is thus more restrictive than a cellular automaton, whose cells may take on more than two values. On the other hand, the nodes of a binary network need not be arranged in a regular grid but can be interconnected in more complex ways.

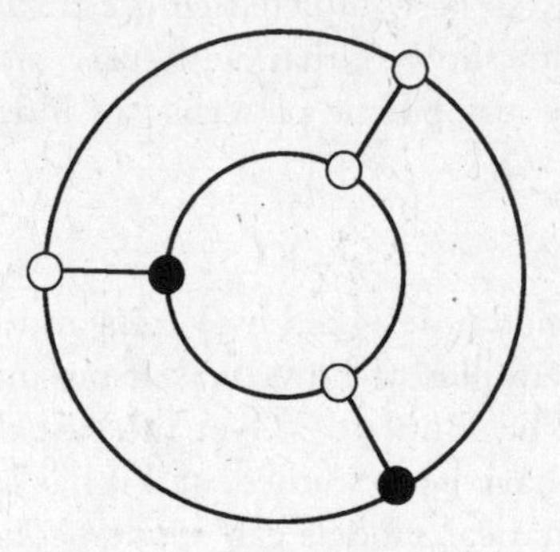

Figure 9-2
A simple binary network.

Binary networks are also called "Boolean networks" after the English mathematician George Boole, who used binary ("yes-no") operations in the mid–nineteenth century to develop a symbolic logic known as Boolean algebra. Figure 9-2 shows a simple binary, or Boolean, network with six nodes, each connected to three neighbors, with two nodes being ON (drawn in black) and four being OFF (drawn in white).

As in a cellular automaton, the pattern of ON-OFF nodes in a binary network changes in discrete steps. The nodes are coupled to one another in such a way that the value of each node is determined by the prior values of neighboring nodes according to some "switching rule." For example, for the network pictured in figure 9-2 we may choose the following switching rule: A node will be ON at the next step if at least two of its neighbors are ON at this step, and OFF in all other cases.

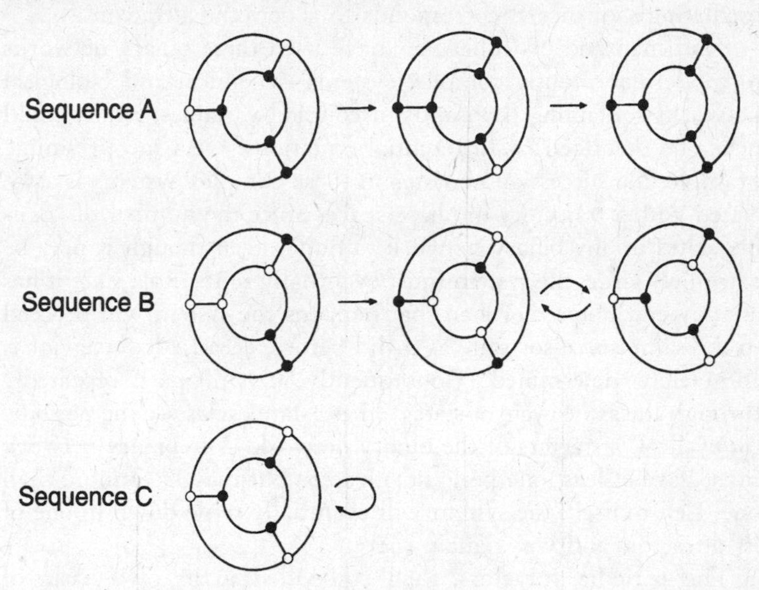

Figure 9-3
Three sequences of states in binary network.

Figure 9-3 shows three sequences generated by this rule. We see that sequence A reaches a stable pattern with all the nodes ON after two steps; sequence B takes one step and then oscillates between two complementary patterns; while the pattern C is stable from the start, reproducing itself at every step. To analyze sequences like these mathematically, each pattern, or state, of the network is defined by six binary (ON-OFF) variables. At each step the system passes from a definite state to a specific successor state, determined completely by the switching rule.

As in systems described by differential equations, each state can be pictured as a point in a six-dimensional phase space.[15] As the network changes step by step from one state to the next, the succession of states traces a trajectory in that phase space. The concept of attractors is used to classify the trajectories of different sequences. Thus in our example the sequence A, which moves toward a stable state, is associated with a point attractor, while the oscillating sequence B corresponds to a periodic attractor.

Kauffman and his colleagues have used these binary networks to model enormously complex systems—chemical and biological networks containing thousands of coupled variables, which could never be described by differential equations.[16] As in our simple example, the succession of states in these complex systems is associated with a trajectory in phase space. Since the number of possible states in any binary network is finite, even though it may be extremely large, the system must eventually return to a state it has already encountered. When that happens the system will proceed to the same successor state as it did before, because its behavior is completely determined. Consequently it will pass repeatedly through the same cycle of states. These state cycles are the periodic (or cyclical) attractors of the binary network. Any binary network must have at least one periodic attractor but may have more than one. Left to itself, the system will eventually settle down to one of its attractors and will remain there.

The periodic attractors, each embedded in its own basin of attraction, are the most important mathematical features of binary networks. Extensive research has shown that a wide variety of living systems—including genetic networks, immune systems, neural networks, organ systems, and ecosystems—can be represented by binary networks exhibiting several alternative attractors.[17]

The different state cycles in a binary network may vary greatly in length. In some networks they can be enormously long, increasing exponentially as the number of nodes increases. Kauffman has defined the attractors of those enormously long cycles, which involve billions and billions of different states, as "chaotic," since their length, for all practical purposes, is infinite.

The detailed analysis of large binary networks in terms of their attractors confirmed what the cyberneticists had already discovered in the 1940s. Although some networks are chaotic, involving seemingly random sequences and infinitely long attractors, others generate small attractors corresponding to patterns of high order. Thus the study of binary networks provides yet another perspective on the phenomenon of self-organization. Networks coordinating the mutual activities of thousands of elements may exhibit vastly ordered dynamics.

At the Edge of Chaos

To investigate the exact relationship between order and chaos in these models, Kauffman examined many complex binary networks and a variety of switching rules, including networks in which the number of "inputs," or links, is different for different nodes. He found that the behavior of these complex webs can be summarized in terms of two parameters: *N,* the number of nodes in the network, and *K,* the average number of inputs to each node. For values of *K* above two—that is, for multiply interconnected networks—the behavior is chaotic, but as *K* gets smaller and approaches two, order crystallizes. Alternatively, order can also emerge at larger values of *K* if the switching rules are "biased"—for example, if there are more possibilities for ON than for OFF.

Detailed studies of the transition from chaos to order have shown that binary networks develop a "frozen core" of elements as the value of *K* approaches two. These are nodes that remain in the same configuration, either ON or OFF, as the system goes through the state cycle. As *K* comes even closer to two, the frozen core creates "walls of constancy" that grow across the entire system, partitioning the network into separate islands of changing elements. These islands are functionally isolated. Changes in the behavior of one island cannot pass through the frozen core to other islands. If *K* decreases further, the islands, too, become frozen; the periodic attractor turns into a point attractor, and the entire network reaches a stable, frozen pattern.

Thus complex binary networks exhibit three broad regimes of behavior: an ordered regime with frozen components, a chaotic regime with no frozen components, and a boundary region between order and chaos where frozen components just begin to "melt." Kauffman's central hypothesis is that living systems exist in that boundary region near the "edge of chaos." He argues that deep in the ordered regime the islands of activity would be too small and isolated for complex behavior to propagate across the system. Deep in the chaotic regime, on the other hand, the system would be too sensitive to small perturbations to maintain its organization. Thus natural selection may favor and sustain living systems "at the edge of chaos," because these may be best able to coordinate complex and flexible behavior, best able to adapt and evolve.

To test his hypothesis, Kauffman applied his model to the genetic networks in living organisms and was able to derive from it several surprising and rather accurate predictions.[18] The great achievements of molecular biology, often described as "the cracking of the genetic code," have made us think of the strands of genes in the DNA as some kind of biochemical computer executing a "genetic program." However, recent research has increasingly shown that this way of thinking is quite erroneous. In fact, it is as inadequate as the metaphor of the brain as an information-processing computer.[19] The complete set of genes in an organism, the so-called genome, forms a vast interconnected network, rich in feedback loops, in which genes directly and indirectly regulate each other's activities. In the words of Francisco Varela, "The genome is not a linear array of independent genes (manifesting as traits) but a highly interwoven network of multiple reciprocal effects mediated through repressors and derepressors, exons and introns, jumping genes, and even structural proteins."[20]

When Stuart Kauffman began to study this complex genetic web, he noticed that each gene in the network is directly regulated by only a few other genes. Moreover, it has been known since the 1960s that the activity of genes, like that of neurons, can be modeled in terms of binary ON-OFF values. Therefore, Kauffman reasoned, binary networks should be appropriate models for genomes. Indeed, this turned out to be the case.

A genome, then, is modeled by a binary network "at the edge of chaos"—that is, a network with a frozen core and separate islands of changing nodes. It will have a relatively small number of state cycles, represented in phase space by periodic attractors embedded in separate basins of attraction. Such a system can undergo two kinds of perturbations. A "minimal" perturbation is an accidental temporary flipping of a binary element into its opposite state. It turns out that each state cycle of the model is remarkably stable under those minimal perturbations. The changes triggered by the perturbation remain confined to a particular island of activity, and after a while the network typically returns to the original state cycle. In other words, the model exhibits the property of homeostasis, which is characteristic of all living systems.

The other kind of perturbation is a permanent structural change in the network—for example, a change in the pattern of connections or in a switching rule—that corresponds to a mutation in the genetic system. Most of these structural perturbations, too, change the behavior of the edge-of-chaos network only slightly. Some, however, may push its trajectory into a different basin of attraction, which results in a new state cycle and thus a new recurrent pattern of behavior. Kauffman sees this as a plausible model for evolutionary adaptation:

> Networks on the boundary between order and chaos may have the flexibility to adapt rapidly and successfully through the accumulation of useful variations. In such poised systems, most mutations have small consequences because of the systems' homeostatic nature. A few mutations, however, cause larger cascades of change. Poised systems will therefore typically adapt to a changing environment gradually, but if necessary, they can occasionally change rapidly.[21]

Another set of impressive explanatory features in Kauffman's model concerns the phenomenon of cell differentiation in the development of living organisms. It is well-known that all cell types in an organism, in spite of their very different shapes and functions, contain roughly the same genetic instructions. Developmental biologists have concluded from this fact that cell types differ from one another not because they contain different genes, but because the genes that are *active* in them differ. In other words, the structure of a genetic network is the same in all cells, but the patterns of genetic activity are different; and since different patterns of genetic activity correspond to different state cycles in the binary network, Kauffman suggests that the different cell types may correspond to different state cycles and, accordingly, to different attractors.

This "attractor model" of cell differentiation leads to several interesting predictions.[22] Each cell in the human body contains about 100,000 genes. In a binary network of that size, the possibilities of different patterns of gene expression are astronomical. However, the number of attractors in such a network at the edge of chaos is approximately equal to the square root of the number of its elements. Therefore a network of 100,000 genes should express itself in about 317 different cell types. This number, derived from very general features of Kauffman's model, comes remarkably close to the 254 distinct cell types identified in humans.

Kauffman has also tested his attractor model with predictions of the number of cell types for various other species and has found that those, too, seem to be related to the number of genes. Figure 9-4 shows his results for several species.[23] The number of cell types and the number of attractors of the corresponding binary networks are seen to rise, more or less in parallel, with the number of genes.

Another two predictions of Kauffman's attractor model concern the stability of cell types. Since the frozen core of the binary network is identical for all attractors, all cell types in an organism should express mostly the same set of genes and should differ by the expressions of only a small percentage of genes. This is indeed the case for all living organisms.

The attractor model also suggests that new cell types are created in the process of development by pushing the system from one basin of attraction into another. Since each basin of attraction has only a few adjacent basins, any single cell type should differentiate by following pathways to its few immediate neighbors, from them to a few additional neighbors, and so on, until the full set of cell types has been created. In other words, cell differentiation should occur along successive branching pathways. Indeed, it is common

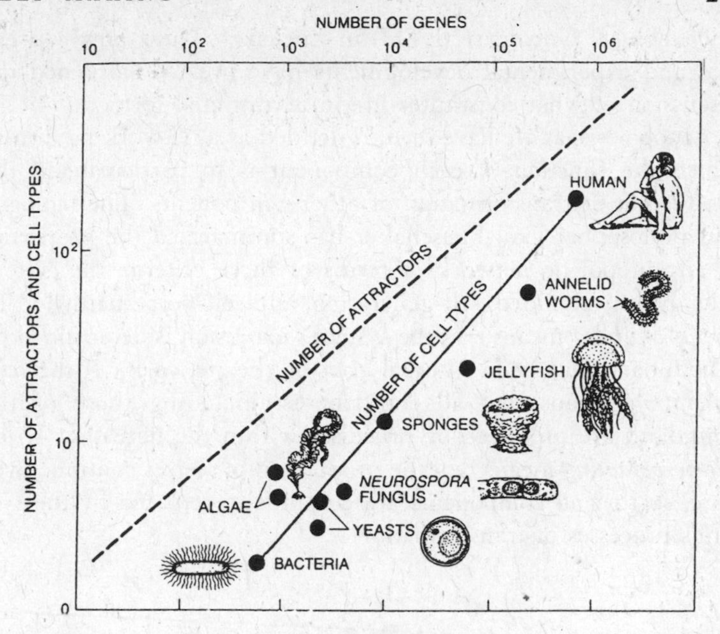

Figure 9-4
Relationships among the number of genes, cell types, and attractors in the corresponding binary networks for different species.

knowledge among biologists that for almost six hundred million years all cell differentiation in multicellular organisms has been organized along such a pattern.

Life in Its Minimal Form

In addition to developing computer simulations of various self-organizing networks—both autopoietic and nonautopoietic—biologists and chemists have also succeeded, more recently, in synthesizing chemical autopoietic systems in the laboratory. This possibility was suggested on theoretical grounds by Francisco Varela and Pier Luigi Luisi in 1989 and was subsequently realized in two kinds of experiments by Luisi and his colleagues at the Swiss

Polytechnical University (ETH) in Zurich.[24] These new conceptual and experimental developments have greatly sharpened the discussion of what constitutes life in its minimal form.

Autopoiesis, as we have seen, is defined as a network pattern in which the function of each component is to participate in the production or transformation of other components. The biologist and philosopher Gail Fleischaker has summarized the properties of an autopoietic network in terms of three criteria: the system must be self-bounded, self-generating, and self-perpetuating.[25] To be *self-bounded* means that the system's extension is determined by a boundary that is an integral part of the network. To be *self-generating* means that all components, including those of the boundary, are produced by processes within the network. To be *self-perpetuating* means that the production processes continue over time, so that all components are continually replaced by the system's processes of transformation.

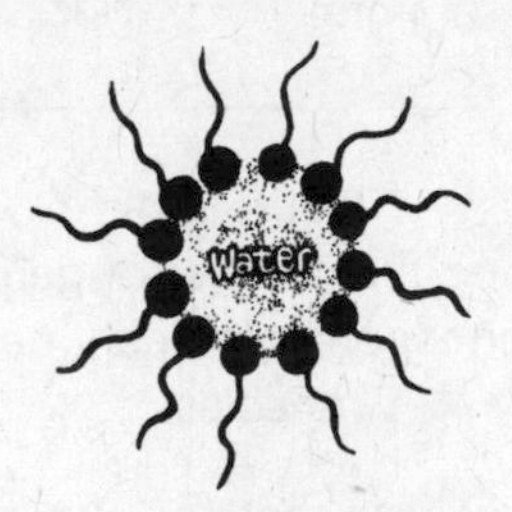

Figure 9-5
Basic Shape of a "Micelle" Droplet

Even though the bacterial cell is the simplest autopoietic system found in nature, the recent ETH experiments showed that chemical structures satisfying the criteria for autopoietic organization can be produced in the laboratory. The first of these structures, suggested by Luisi and Varela in their theoretical paper, is known to chemists as a "micelle." It is basically a water droplet surrounded by a thin layer of tadpole-shaped molecules with water-attracting "heads" and water-repelling "tails" (see figure 9-5).

Under special circumstances such a droplet may host chemical reactions producing certain components, which organize themselves into the very boundary molecules that build the structure and provide the conditions for the reactions to take place. Thus a simple chemical autopoietic system is created. As in Varela's computer simulation, the reactions are enclosed by a boundary assembled from the very products of the reactions.

After this first example of autopoietic chemistry, the researchers at ETH succeeded in creating another type of chemical structure that is even more relevant to cellular processes, because its main ingredients—so-called fatty acids—are thought to have been the material for primordial cell walls. The experiments consisted in producing spherical water droplets surrounded by shells of those fatty substances, which have the typical semipermeable structure of biological membranes (but without their protein components) and generate catalytic loops resulting in an autopoietic system. The researchers who carried out the experiments speculate that these kinds of systems may have been the first closed self-reproducing chemical structures before the evolution of the bacterial cell. If this is true, it would mean that scientists have now succeeded in re-creating the first minimal forms of life.

Organisms and Societies

Most of the research in the theory of autopoiesis, so far, has been concerned with minimal autopoietic systems—simple cells, computer simulations, and the recently discovered autopoietic chemical structures. Much less work has been done on studying the autopoiesis of multicellular organisms, ecosystems, and social systems. Current ideas about the network patterns in those living systems are therefore still rather speculative.[26]

All living systems are networks of smaller components, and the web of life as a whole is a multilayered structure of living systems nesting within other living systems—networks within networks. Organisms are aggregates of autonomous but closely coupled cells; populations are networks of autonomous organisms belonging to a single species; and ecosystems are webs of organisms, both single celled and multicellular, belonging to many different species.

What is common to all these living systems is that their smallest living components are always cells, and therefore we can confidently say that all living systems, ultimately, are autopoietic. However, it is also interesting to ask whether the larger systems formed by those autopoietic cells—the organisms, societies, and ecosystems—are in themselves autopoietic networks.

In their book *The Tree of Knowledge,* Maturana and Varela argue that our current knowledge about the details of the metabolic pathways in organisms and ecosystems is not sufficient to give a clear answer, and hence they leave the question open:

> What we can say is that [multicellular systems] have *operational closure* in their organization: their identity is specified by a network of dynamic processes whose effects do not leave the network. But regarding the explicit form of that organization, we shall not speak further.[27]

The authors then go on to point out that the three types of multicellular living systems—organisms, ecosystems, and societies—differ greatly in the degrees of autonomy of their components. In organisms the cellular components have a minimal degree of independent existence, while the components of human societies, individual human beings, have a maximum degree of autonomy, enjoying many dimensions of independent existence. Animal societies and ecosystems occupy various places between those two extremes.

Human societies are a special case because of the crucial role of language, which Maturana has identified as the critical phenomenon in the development of human consciousness and culture.[28] While the cohesion of social insects is based on the exchange of chemicals between the individuals, the social unity of human societies is based on the exchange of language.

The components of an organism exist for the organism's functioning, but human social systems exist also *for their components,* the individual human beings. Thus, in the words of Maturana and Varela:

> The organism restricts the individual creativity of its component unities, as these unities exist for that organism. The human social

system amplifies the individual creativity of its components, as that system exists for these components.[29]

Organisms and human societies are therefore very different types of living systems. Totalitarian political regimes have often severely restricted the autonomy of their members and, in doing so, have depersonalized and dehumanized them. Thus fascist societies function more like organisms, and it is not a coincidence that dictatorships have often been fond of using the metaphor of society as a living organism.

Autopoiesis in the Social Domain

The question of whether human social systems can be described as autopoietic has been discussed quite extensively, and different authors have proposed various answers.[30] The central problem is that autopoiesis has been defined precisely only for systems in physical space and for computer simulations in mathematical spaces. Because of the "inner world" of concepts, ideas, and symbols that arises with human thought, consciousness, and language, human social systems exist not only in the physical domain but also in a symbolic social domain.

Thus a human family can be described as a biological system, defined by certain blood relations, but also as a "conceptual system," defined by certain roles and relationships that may or may not coincide with any blood relationships among its members. These roles depend on social convention and may vary considerably in different periods of time and different cultures. For example, in contemporary Western culture the role of "father" may be fulfilled by the biological father, a foster father, a stepfather, an uncle, or an older brother. In other words, these roles are not objective features of the family system but are flexible and continually renegotiated social constructs.[31]

While behavior in the physical domain is governed by cause and effect, the so-called "laws of nature," behavior in the social domain is governed by rules generated by the social system and often codified into law. The crucial difference is that social rules can be broken, but natural laws cannot. Human beings can choose whether and how to obey a social rule; molecules cannot choose whether or not they should interact.[32]

Given the simultaneous existence of social systems in two domains, the physical and the social, is it meaningful to apply the concept of autopoiesis to them at all, and if so, in which domain should it be applied?

After leaving this question open in their book, Maturana and Varela have expressed separate and slightly different views. Maturana does not see human social systems as being autopoietic, but rather as the medium in which human beings realize their biological autopoiesis through "languaging."[33] Varela argues that the concept of a network of production processes, which is at the very core of the definition of autopoiesis, may not be applicable beyond the physical domain, but that a broader concept of "organizational closure" can be defined for social systems. This broader concept is similar to that of autopoiesis but does not specify processes of production.[34] Autopoiesis, in Varela's view, can be seen as a special case of organizational closure, manifest at the cellular level and in certain chemical systems.

Other authors have asserted that an autopoietic social network *can* be defined if the description of human social systems remains entirely within the social domain. This school of thought was pioneered in Germany by sociologist Niklas Luhmann, who has developed the concept of social autopoiesis in considerable detail. Luhmann's central point is to identify the social processes of the autopoietic network as processes of communication:

> Social systems use communication as their particular mode of autopoietic reproduction. Their elements are communications that are . . . produced and reproduced by a network of communications and that cannot exist outside of such a network.[35]

A family system, for example, can be defined as a network of conversations exhibiting inherent circularities. The results of conversations give rise to further conversations, so that self-amplifying feedback loops are formed. The closure of the network results in a shared system of beliefs, explanations, and values—a context of meaning—that is continually sustained by further conversations.

The communicative acts of the network of conversations include the "self-production" of the roles by which the various family members are defined and of the family system's boundary. Since all these processes take place in the symbolic social domain, the boundary cannot be a physical boundary. It is a boundary of expectations, confidentiality, loyalty, and so on. Both the family roles and boundaries are continually maintained and renegotiated by the autopoietic network of conversations.

The Gaia System

Whereas the debate on autopoiesis in social systems has been very lively over the past few years, it is surprising that there has been almost total silence on the question of autopoiesis in ecosystems. One would have to agree with Maturana and Varela that the many pathways and processes in an ecosystem are not yet known in sufficient detail to decide whether such an ecological network can be described as autopoietic. However, it would certainly be as interesting to begin discussions on autopoiesis with ecologists as it has been with social scientists.

To begin with, we can say that a function of all components in a food web is to transform other components within the same web. As plants take up inorganic matter from their environment to produce organic compounds, and as these compounds are passed on through the ecosystem to serve as food for the production of more complex structures, the entire network regulates itself through multiple feedback loops.[36] Individual components of the food web continually die, to be decomposed and replaced by the network's own processes of transformation. Whether this is sufficient to define an ecosystem as autopoietic remains to be seen and depends, among other things, on a clear understanding of the system's boundary.

When we shift our perception from ecosystems to the planet as a whole, we encounter a global network of processes of production and transformation, which has been described in some detail in the Gaia theory by James Lovelock and Lynn Margulis.[37] In fact, today there may be more evidence for the autopoietic nature of the Gaia system than for that of ecosystems.

The planetary system operates on a very large scale in space and also involves long time scales. It is thus not so easy to think of Gaia as being alive in a concrete manner. Is the whole planet alive or just certain parts? And if the latter, which parts? To help us picture Gaia as a living system, Lovelock has suggested a tree as an analogy.[38] As the tree grows, there is only a thin layer of living cells around its perimeter, just beneath the bark. All the wood inside, more than 97 percent of the tree, is dead. Similarly, the Earth is covered with a thin layer of living organisms—the biosphere—reaching down into the ocean about five to six miles and up into the atmosphere about the same distance. So the living part of Gaia is but a thin film around the globe. If the planet is represented by a globe the size of a basketball with the oceans and countries painted on it, the thickness of the biosphere would be just about the thickness of the paint!

Just as the bark of a tree protects the tree's thin layer of living tissue from damage, life on Earth is surrounded by the protective layer of atmosphere, which shields us from ultraviolet light and other harmful influences and keeps the planet's temperature just right for life to flourish. Neither the atmosphere above us nor the rocks below us are alive, but both have been shaped and transformed considerably by living organisms, just like the bark and the wood of the tree. Outer space and the Earth's interior are both part of Gaia's environment.

To see whether the Gaia system can indeed be described as an autopoietic network, let us apply the three criteria proposed by Gail Fleischaker.[39] Gaia is definitely *self-bounded* at least as far as the outer boundary, the atmosphere, is concerned. According to the Gaia theory, the Earth's atmosphere is created, transformed, and maintained by the biosphere's metabolic processes. Bacteria play a crucial role in these processes, influencing the rate of chemical reactions and thus acting as the biological equivalent of the enzymes in a cell.[40] The atmosphere is semipermeable, like a cell membrane, and forms an integral part of the planetary network.

For example, it created the protective greenhouse in which early life on the planet was able to unfold three billion years ago, even though the sun was then 25 percent less luminous than it is now.[41]

The Gaia system is also clearly *self-generating*. The planetary metabolism converts inorganic substances into organic, living matter and back into soil, oceans, and air. All components of the Gaian network, including those of its atmospheric boundary, are produced by processes within the network.

A key characteristic of Gaia is the complex interweaving of living and nonliving systems within a single web. This results in feedback loops of vastly differing scales. Rock cycles, for example, extend over hundreds of millions of years, while the organisms associated with them have very short life spans. In the metaphor of Stephan Harding, ecologist and collaborator of James Lovelock: "Living beings come out of rocks and go back into rocks."[42]

Finally, the Gaia system is evidently *self-perpetuating*. The components of the oceans, soil, and air, as well as all the organisms of the biosphere, are continually replaced by the planetary processes of production and transformation. It seems, then, that the case for Gaia being an autopoietic network is very strong. Indeed, Lynn Margulis, coauthor of the Gaia theory, asserts confidently: "There is little doubt that the planetary patina—including ourselves—is autopoietic."[43]

The confidence of Lynn Margulis in the idea of a planetary autopoietic web stems from three decades of pioneering work in microbiology. To understand the complexity, diversity, and self-organizing capabilities of the Gaian network, an understanding of the microcosm—the nature, extension, metabolism, and evolution of microorganisms—is absolutely essential. Margulis has not only contributed a great deal to that understanding within the scientific community but has also been able, in collaboration with Dorion Sagan, to explain her radical discoveries in clear and engaging language to the lay reader.[44]

Life on Earth began around 3.5 billion years ago, and for the first 2.0 billion years the living world consisted entirely of microorganisms. During the first billion years of evolution, bacteria—the most basic forms of life—covered the planet with an intricate web of metabolic processes and began to regulate the temperature and chemical composition of the atmosphere so that it became conducive to the evolution of higher forms of life.[45]

Plants, animals, and humans are latecomers to the Earth, having emerged from the microcosm less than one billion years ago. Even today the visible living organisms function only because of their well-developed connections with the bacterial web of life. "Far from leaving microorganisms behind on an evolutionary 'ladder,' " writes Margulis, "we are both surrounded by them and composed of them. . . . [We have to] think of ourselves and our environment as an evolutionary mosaic of microcosmic life."[46]

During life's long evolutionary history, over 99 percent of all species that ever existed have become extinct, but the planetary web of bacteria has survived, continuing to regulate the conditions for life on Earth as it has for the past three billion years. According to Margulis, the concept of a planetary autopoietic network is justified because all life is embedded in a self-organizing web of bacteria, involving elaborate networks of sensory and control systems that we are only beginning to recognize. Myriad bacteria, living in the soil, the rocks, and the oceans, as well as inside all plants, animals, and humans, continually regulate life on Earth: "It is the growth, metabolism, and gas-exchanging properties of microbes . . . that form the complex physical and chemical feedback systems which modulate the biosphere in which we live."[47]

The Universe at Large

Reflecting on the planet as a living being, one is naturally led to ask questions about systems of even larger scales. Is the solar system an autopoietic network? The galaxy? And what about the universe as a whole? Is the universe alive?

Regarding the solar system, we can say with some confidence that it does not appear to be a living system. Indeed, it was the striking difference between the Earth and all other planets in the solar system that led Lovelock to formulate the Gaia hypothesis. As far as our galaxy, the Milky Way, is concerned, we are nowhere near to having the data necessary to entertain the question of whether it is alive, and when we shift our perspective to the universe as a whole we also reach the limits of conceptualization.

For many people, including myself, it is philosophically and spiritually more satisfying to assume that the cosmos as a whole is alive, rather than thinking of life on Earth existing within a lifeless universe. Within the framework of science, however, we cannot—or, at least, not yet—make such statements. If we apply our scientific criteria for life to the entire universe, we encounter serious conceptual difficulties.

Living systems are defined as being open to a constant flow of energy and matter. But how can we think of the universe, which by definition includes everything, as an open system? The question does not seem to make any more sense than to ask what happened before the Big Bang. In the words of the renowned astronomer Sir Bernard Lovell:

> There we reach the great barrier of thought. . . . I feel as though I've suddenly driven into a great fog barrier where the familiar world has disappeared.[48]

One thing we *can* say about the universe is that the potential for life exists in abundance throughout the cosmos. Research over the last few decades has provided a fairly clear picture of the geological and chemical features on the early Earth that made life possible. We have begun to understand how more and more complex chemical systems developed and how they formed catalytic cycles that, eventually, evolved into autopoietic systems.[49]

Observing the universe at large, and our galaxy in particular, astronomers have discovered that the characteristic chemical components found in all life are present in abundance. For life to emerge from these compounds, a delicate balance of temperatures, atmospheric pressures, water content, and so on is required. During the long evolution of the galaxy, it is likely that this balance was achieved on many planets in the billions of planetary systems the galaxy contains.

Even in our solar system, both Venus and Mars probably had oceans in their early history in which life could have emerged.[50] But Venus was too close to the sun for a slow pace of evolution. Its oceans evaporated, and eventually the hydrogen was split off from the water molecules by powerful ultraviolet radiation and escaped into space. We do not know how Mars lost its water; we only know that it did. Lovelock speculates that perhaps Mars had life in the early stages and lost it in some catastrophic event, or that hydrogen escaped faster than on the early Earth because of the much weaker force of gravity on Mars.

Be that as it may, it seems that life "almost" evolved on Mars and that in all likelihood it did evolve and is flourishing on millions of other planets throughout the universe. Thus even though the concept of the universe as a whole being alive is problematic within the framework of present-day science, we can say with confidence that life is probably present in great abundance throughout the cosmos.

Structural Coupling

Wherever we see life, from bacteria to large-scale ecosystems, we observe networks with components that interact with one another in such a way that the entire network regulates and organizes itself. Since these components, except for those in cellular networks, are themselves living systems, a realistic picture of autopoietic networks must include a description of how living systems interact with one another and, more generally, with their environment. Indeed, such a description is an integral part of the theory of autopoiesis developed by Maturana and Varela.

The central characteristic of an autopoietic system is that it undergoes continual structural changes while preserving its weblike pattern of organization. The components of the network continually produce and transform one another, and they do so in two distinct ways. One type of structural changes are changes of self-renewal. Every living organism continually renews itself, cells breaking down and building up structures, tissues and organs replacing their cells in continual cycles. In spite of this ongoing change, the organism maintains its overall identity, or pattern of organization.

Many of these cyclical changes occur much faster than one

would imagine. For example, our pancreas replaces most of its cells every twenty-four hours, the cells of our stomach lining are reproduced every three days, our white blood cells are renewed in ten days, and 98 percent of the protein in our brain is turned over in less than one month. Even more amazing, our skin replaces its cells at the rate of one hundred thousand cells per minute. In fact, most of the dust in our homes consists of dead skin cells.

The second type of structural changes in a living system are changes in which new structures are created—new connections in the autopoietic network. These changes of the second type—developmental rather than cyclical—also take place continually, either as a consequence of environmental influences or as a result of the system's internal dynamics. According to the theory of autopoiesis, a living system interacts with its environment through "structural coupling," that is, through recurrent interactions, each of which triggers structural changes in the system. For example, a cell membrane continually incorporates substances from its environment into the cell's metabolic processes. An organism's nervous system changes its connectivity with every sense perception. These living systems are autonomous, however. The environment only triggers the structural changes; it does not specify or direct them.[51]

Structural coupling, as defined by Maturana and Varela, establishes a clear difference between the ways living and nonliving systems interact with their environments. Kicking a stone and kicking a dog are two very different stories, as Gregory Bateson was fond of pointing out. The stone will *react* to the kick according to a linear chain of cause and effect. Its behavior can be calculated by applying the basic laws of Newtonian mechanics. The dog will *respond* with structural changes according to its own nature and (nonlinear) pattern of organization. The resulting behavior is generally unpredictable.

As a living organism responds to environmental influences with structural changes, these changes will in turn alter its future behavior. In other words, a structurally coupled system is a learning system. As long as it remains alive, a living organism will couple structurally to its environment. Its continual structural changes in response to the environment—and consequently its continuing adaptation, learning, and development—are key characteristics of the behavior of living beings. Because of its structural coupling, we call the behavior of an animal intelligent but would not apply that term to the behavior of a rock.

Development and Evolution

As it keeps interacting with its environment, a living organism will undergo a sequence of structural changes, and over time it will form its own, individual pathway of structural coupling. At any point on this pathway, the structure of the organism is a record of previous structural changes and thus of previous interactions. Living structure is always a record of previous development, and ontogeny—the course of development of an individual organism—is the organism's history of structural changes.

Now, since an organism's structure at any point in its development is a record of its previous structural changes, and since each structural change influences the organism's future behavior, this implies that the behavior of the living organism is determined by its structure. Thus a living system is determined in different ways by its pattern of organization and its structure. The pattern of organization determines the system's identity (its essential characteristics); the structure, formed by a sequence of structural changes, determines the system's behavior. In Maturana's terminology the behavior of living systems is "structure-determined."

This concept of structural determinism sheds new light on the age-old philosophical debate about freedom and determinism. According to Maturana, the behavior of a living organism is determined. However, rather than being determined by outside forces, it is determined by the organism's own structure—a structure formed by a succession of autonomous structural changes. Thus the behavior of the living organism is both determined and free.

Moreover, the fact that the behavior is structure-determined does not mean that it is predictable. The organism's structure merely "conditions the course of its interactions and restricts the structural changes that the interactions may trigger in it."[52] For example, when a living system reaches a bifurcation point, as described by Prigogine, its history of structural coupling will determine the new pathways that become available, but which pathway the system will take remains unpredictable.

Like Prigogine's theory of dissipative structures, the theory of autopoiesis shows that creativity—the generation of configurations that are constantly new—is a key property of all living systems. A special form of this creativity is the generation of diversity through reproduction, from simple cell division to the highly complex dance of sexual reproduction. For most living organisms ontogeny is not a linear path of development but a cycle, and reproduction is a vital step in that cycle.

Billions of years ago the combined abilities of living systems to reproduce and create novelty led naturally to biological evolution—a creative unfolding of life that has continued in an uninterrupted process ever since. From the most archaic and simple forms of life to the most intricate and complex contemporary forms, life has unfolded in a continual dance without ever breaking the basic pattern of its autopoietic networks.

10

The Unfolding of Life

One of the most rewarding features of the emerging theory of living systems is the new understanding of evolution it implies. Rather than seeing evolution as the result of random mutations and natural selection, we are beginning to recognize the creative unfolding of life in forms of ever-increasing diversity and complexity as an inherent characteristic of all living systems. Although mutation and natural selection are still acknowledged as important aspects of biological evolution, the central focus is on creativity, on life's constant reaching out into novelty.

To understand the fundamental difference between the old and new views of evolution, it will be useful to briefly review the history of evolutionary thought.

Darwinism and Neo-Darwinism

The first theory of evolution was formulated at the beginning of the nineteenth century by Jean Baptiste Lamarck, a self-taught naturalist who coined the term "biology" and made extensive studies in botany and zoology. Lamarck observed that animals changed under environmental pressure, and he believed that they could pass on these changes to their offspring. This passing on of acquired characteristics was for him the main mechanism of evolution.

Although it turned out that Lamarck was wrong in that respect, his recognition of the phenomenon of evolution—the emergence of new biological structures in the history of species—was a revolutionary insight that profoundly affected all subsequent scientific thought. In particular Lamarck had a strong influence on Charles Darwin, who started his scientific career as a geologist but became interested in biology during his famous expedition to the Galápagos Islands. His careful observations of the island fauna stimulated Darwin to speculate about the effect of geographical isolation on the formation of species and led him, eventually, to the formulation of his theory of evolution.

Darwin published his theory in 1859 in his monumental work *On the Origin of Species* and completed it twelve years later with *The Descent of Man,* in which the concept of evolutionary transformation of one species into another is extended to include human beings. Darwin based his theory on two fundamental ideas—chance variation, later to be called random mutation, and natural selection.

At the center of Darwinian thought stands the insight that all living organisms are related by common ancestry. All forms of life have emerged from that ancestry by a continuous process of variations throughout billions of years of geological history. In this evolutionary process many more variations are produced than can possibly survive, and thus many individuals are weeded out by natural selection, as some variants outgrow and outreproduce others.

These basic ideas are well documented today, supported by vast amounts of evidence from biology, biochemistry, and the fossil record, and all serious scientists are in complete agreement with them. The differences between the classical theory of evolution and the emerging new theory center around the question of the *dynamics* of evolution—the mechanisms through which evolutionary changes take place.

Darwin's own concept of chance variations was based on an assumption that was common to nineteenth-century views of heredity. It was assumed that the biological characteristics of an individual represented a "blend" of those of its parents, with both parents contributing more or less equal parts to the mixture. This meant that an offspring of a parent with a useful chance variation would inherit only 50 percent of the new characteristic and would be able to pass on only 25 percent of it to the next generation. Thus the new characteristic would be diluted rapidly, with very little chance of establishing itself through natural selection. Darwin himself recognized that this was a serious flaw in his theory for which he had no remedy.

It is ironic that the solution to Darwin's problem was discovered by Gregor Mendel, an Austrian monk and amateur botanist, only a few years after the publication of the Darwinian theory but was ignored during Mendel's lifetime and brought to light again only at the turn of the century, many years after Mendel's death. From his careful experiments with garden peas, Mendel deduced that there were "units of heredity"—later to be called genes—that did not blend in the process of reproduction but were transmitted from generation to generation without changing their identity. With this discovery it could be assumed that random mutations of genes would not disappear within a few generations but would be preserved, to be either reinforced or eliminated by natural selection.

Mendel's discovery not only played a decisive role in establishing the Darwinian theory of evolution but also opened up a whole new field of research—the study of heredity through the investigation of the chemical and physical nature of genes.[1] A British biologist, William Bateson, a fervent advocate and popularizer of Mendel's work, called this new field "genetics" at the beginning of the century. He also named his youngest son, Gregory, in Mendel's honor.

The combination of Darwin's idea of gradual evolutionary changes with Mendel's discovery of genetic stability resulted in the synthesis known as neo-Darwinism, which is taught today as the established theory of evolution in biology departments around the world. According to the neo-Darwinist theory, all evolutionary variation results from random mutation—that is, from random genetic changes—followed by natural selection. For example, if an animal species needs thick fur to survive in a cold climate, it will not respond to this need by growing fur but will instead develop all sorts of random genetic changes, and those animals whose changes happen to result in thick fur will survive to produce more offspring. Thus, in the words of geneticist Jacques Monod, "Chance alone is at the source of every innovation, of all creation in the biosphere."[2]

In the view of Lynn Margulis, neo-Darwinism is fundamentally flawed, not only because it is based on reductionist concepts that are now outdated, but also because it was formulated in an inappropriate mathematical language. "The language of life is not ordinary arithmetic and algebra," argues Margulis, "the language of life is chemistry. The practicing neo-Darwinists lack relevant knowledge in, for example, microbiology, cell biology, biochemistry . . . and microbial ecology."[3]

One reason why today's leading evolutionists lack the appropriate language to describe evolutionary change, according to Margulis, is that most of them come out of the zoological tradition and thus are used to dealing with only a small, relatively recent part of evolutionary history. Current research in microbiology indicates strongly that the major avenues for evolution's creativity were developed long before animals appeared on the scene.[4]

The central conceptual problem of neo-Darwinism seems to be its reductionist conception of the genome, the collection of an organism's genes. The great achievements of molecular biology, often described as "the cracking of the genetic code," have resulted in the tendency to picture the genome as a linear array of independent genes, each corresponding to a biological trait.

Research has shown, however, that a single gene may affect a wide range of traits and that, conversely, many separate genes often combine to produce a single trait. It is thus quite mysterious how complex structures, like an eye or a flower, could have evolved through successive mutations of individual genes. Evidently the study of the coordinating and integrating activities of the whole genome is of paramount importance, but this has been hampered severely by the mechanistic outlook of conventional biology. Only very recently have biologists begun to understand the genome of an organism as a highly interwoven network and to study its activities from a systemic perspective.[5]

The Systems View of Evolution

A striking manifestation of genetic wholeness is the now well-documented fact that evolution did not proceed through continuous gradual changes over time, caused by long sequences of successive mutations. The fossil record shows clearly that throughout evolutionary history there have been long periods of stability, or "stasis," without any genetic variation, punctuated by sudden and dramatic transitions. Stable periods of hundreds of thousands of years are quite the norm. Indeed, the human evolutionary adventure began with a million years of stability of the first hominid species, *Australopithecus afarensis.*[6] This new picture, known as "punctuated equilibria," indicates that the sudden transitions were caused by mechanisms quite different from the random mutations of neo-Darwinist theory.

An important aspect of the classical theory of evolution is the idea that in the course of evolutionary change and under the pressure of natural selection, organisms will gradually adapt to their environment until they reach a fit that is good enough for survival and reproduction. In the new systems view, by contrast, evolutionary change is seen as the result of life's inherent tendency to create novelty, which may or may not be accompanied by adaptation to changing environmental conditions.

Accordingly, systems biologists have begun to portray the genome as a self-organizing network capable of spontaneously producing new forms of order. "We must rethink evolutionary biology," writes Stuart Kauffman. "Much of the order we see in organisms may be the direct result not of natural selection but of the natural order selection was privileged to act on. . . . Evolution is not just a tinkering. . . . It is emergent order honored and honed by selection."[7]

A comprehensive new theory of evolution, based on these recent insights, has not yet been formulated. But the models and theories of self-organizing systems discussed in the previous chapters of this book provide the elements for formulating such a theory.[8] Prigogine's theory of dissipative structures shows how complex biochemical systems, operating far from equilibrium, generate catalytic loops that lead to instabilities and can produce new structures of higher order. Manfred Eigen has suggested that similar catalytic cycles may have formed before the emergence of life on Earth, thus initiating a prebiological phase of evolution. Stuart Kauffman has used binary networks as mathematical models of the genetic networks of living organisms and was able to derive several known features of cell differentiation and evolution from these models. Humberto Maturana and Francisco Varela have described the process of evolution in terms of their theory of autopoiesis, seeing the evolutionary history of a species as the history of its structural coupling. And James Lovelock and Lynn Margulis in their Gaia theory have explored the planetary dimensions of the unfolding of life.

The Gaia theory, as well as the earlier work by Lynn Margulis in microbiology, have exposed the fallacy of the narrow Darwinian concept of adaptation. Throughout the living world evolution cannot be limited to the adaptation of organisms to their environment, because the environment itself is shaped by a network of living systems capable of adaptation and creativity. So, which adapts to which? Each to the other—they *coevolve*. As James Lovelock put it:

> So closely coupled is the evolution of living organisms with the evolution of their environment that together they constitute a single evolutionary process.[9]

Thus our focus is shifting from evolution to coevolution—an ongoing dance that proceeds through a subtle interplay of competition and cooperation, creation and mutual adaptation.

Avenues of Creativity

So the driving force of evolution, according to the emerging new theory, is to be found not in the chance events of random mutations, but in life's inherent tendency to create novelty, in the spontaneous emergence of increasing complexity and order. Once this fundamental new insight has been understood, we can then ask: What are the avenues in which evolution's creativity expresses itself?

The answer to this question comes not only from molecular biology, but also, and even more importantly, from microbiology, from the study of the planetary web of the myriad microorganisms that were the only forms of life during the first two billion years of evolution. During those two billion years bacteria continually transformed the Earth's surface and atmosphere and, in so doing, invented all of life's essential biotechnologies, including fermentation, photosynthesis, nitrogen fixation, respiration, and rotary devices for rapid motion.

During the past three decades extensive research in microbiology has revealed three major avenues of evolution.[10] The first, but least important, is the random mutation of genes, the centerpiece of neo-Darwinian theory. Gene mutation is caused by a chance error in the self-replication of DNA, when the two chains of the DNA's double helix separate and each of them serves as a template for the construction of a new complementary chain.[11]

It has been estimated that those chance errors occur at a rate of about one per several hundred million cells in each generation. This frequency does not seem to be sufficient to explain the evolution of the great diversity of life forms, given the well-known fact that most mutations are harmful and only very few result in useful variations.

In the case of bacteria the situation is different, because bacteria divide so rapidly. Fast bacteria can divide about every twenty minutes, so that in principle several billion individual bacteria can be generated from a single cell in less than a day.[12] Because of this enormous rate of reproduction, a single successful bacterial mutant can spread rapidly through its environment, and mutation is indeed an important evolutionary avenue for bacteria.

However, bacteria have developed a second avenue of evolutionary creativity that is vastly more effective than random mutation. They freely pass hereditary traits from one to another in a global exchange network of incredible power and efficiency. Here is how Lynn Margulis and Dorion Sagan describe it:

> Over the past fifty years or so, scientists have observed that [bacteria] routinely and rapidly transfer different bits of genetic material to other individuals. Each bacterium at any given time has the use of accessory genes, visiting from sometimes very different strains, which perform functions that its own DNA may not cover. Some of the genetic bits are recombined with the cell's native genes; others are passed on again. . . . As a result of this ability, all the world's bacteria essentially have access to a single gene pool and hence to the adaptive mechanisms of the entire bacterial kingdom.[13]

This global trading of genes, technically known as DNA recombination, must rank as one of the most astonishing discoveries of modern biology. "If the genetic properties of the microcosm were applied to larger creatures, we would have a science-fiction world," write Margulis and Sagan, "in which green plants could share genes for photosynthesis with nearby mushrooms, or where people could exude perfumes or grow ivory by picking up genes from a rose or a walrus."[14]

The speed with which drug resistance spreads among bacterial communities is dramatic proof that the efficiency of their communications network is vastly superior to that of adaptation through mutations. Bacteria are able to adapt to environmental changes in a few years, where larger organisms would need thousands of years of evolutionary adaptation. Thus microbiology teaches us the sobering lesson that technologies like genetic engineering and a global communications network, which we consider to be advanced achievements of our modern civilization, have been used by the planetary web of bacteria for billions of years to regulate life on Earth.

The constant trading of genes among bacteria results in an amazing variety of genetic structures besides their main strand of DNA. These include the formation of viruses, which are not full autopoietic systems but consist merely of a stretch of DNA or RNA in a protein coating.[15] In fact, Canadian bacteriologist Sorin

Sonea has argued that bacteria, strictly speaking, should not be classified into species, since all of their strains can potentially share hereditary traits and, typically, change up to 15 percent of their genetic material on a daily basis. "A bacterium is not a unicellular organism," writes Sonea; "it is an incomplete cell . . . belonging to different chimeras according to circumstances."[16] In other words, all bacteria are part of a single microcosmic web of life.

Evolution through Symbiosis

Mutation and DNA recombination (the trading of genes) are the two principal avenues for bacterial evolution. But what about the multicellular organisms of all the larger forms of life? If random mutations are not an effective evolutionary mechanism for them, and if they do not trade genes like bacteria, how have the higher forms of life evolved? This question was answered by Lynn Margulis with the discovery of a third, totally unexpected avenue of evolution that has profound implications for all branches of biology.

Microbiologists have known for some time that the most fundamental division among all forms of life is not that between plants and animals, as most people assume, but one between two kinds of cells—cells with and without a cell nucleus. Bacteria, the simplest life forms, do not have cell nuclei and are therefore also called *prokaryotes* ("non-nucleated cells"), whereas all other cells have nuclei and are called *eukaryotes* ("nucleated cells"). All the cells of higher organisms are nucleated, and eukaryotes also appear as single-celled, nonbacterial microorganisms.

In her study of genetics Margulis became intrigued by the fact that not all the genes in a nucleated cell are found inside the cell nucleus:

> We were all taught that the genes were in the nucleus and that the nucleus is the central control of the cell. Early in my study of genetics, I became aware that other genetic systems with different inheritance patterns exist. From the beginning I was curious about those unruly genes that weren't in the nucleus.[17]

As she studied this phenomenon more closely, Margulis found out that nearly all the "unruly genes" are derived from bacteria, and gradually she came to realize that they belong to distinct living organisms, live small cells residing inside larger cells.

Symbiosis, the tendency of different organisms to live in close association with one another and often inside one another (like the bacteria in our intestines), is a widespread and well-known phenomenon. But Margulis went a step further and proposed the hypothesis that long-term symbioses, involving bacteria and other microorganisms living inside larger cells, have led and continue to lead to new forms of life. Margulis published her revolutionary hypothesis first in the mid-1960s and over the years developed it into a full-fledged theory, now known as "symbiogenesis," which sees the creation of new forms of life through permanent symbiotic arrangements as the principal avenue of evolution for all higher organisms.

The most striking evidence for evolution through symbiosis is presented by the so-called mitochondria, the "powerhouses" inside most nucleated cells.[18] These vital parts of all animal and plant cells, which carry out cellular respiration, contain their own genetic material and reproduce independently and at different times from the rest of the cell. Margulis speculates that the mitochondria were originally free-floating bacteria that in ancient times invaded other microorganisms and took up permanent residence inside them. "The merged organisms went on to evolve into more complex oxygen-breathing forms of life," Margulis explains. "Here, then, was an evolutionary mechanism more sudden than mutation: a symbiotic alliance that becomes permanent."[19]

The theory of symbiogenesis implies a radical shift of perception in evolutionary thought. Whereas the conventional theory sees the unfolding of life as a process in which species only diverge from one another, Lynn Margulis claims that the formation of new composite entities through the symbiosis of formerly independent organisms has been the more powerful and more important evolutionary force.

This new view has forced biologists to recognize the vital importance of cooperation in the evolutionary process. While the social Darwinists of the nineteenth century saw only competition in nature—"nature, red in tooth and claw," as the poet Tennyson put it—we are now beginning to see continual cooperation and mutual dependence among all life forms as central aspects of evolution. In the words of Margulis and Sagan, "Life did not take over the globe by combat, but by networking."[20]

The evolutionary unfolding of life over billions of years is a breathtaking story. Driven by the creativity inherent in all living systems, expressed through three distinct avenues—mutations, the trading of genes, and symbioses—and honed by natural selection, the planet's living patina expanded and intensified in forms of ever-increasing diversity. The story is told beautifully by Lynn Margulis and Dorion Sagan in their book *Microcosmos,* on which the following pages are largely based.[21]

There is no evidence of any plan, goal, or purpose in the global evolutionary process and thus no evidence for progress; yet there are recognizable patterns of development. One of these, known as convergence, is the tendency of organisms to evolve similar forms for meeting similar challenges, in spite of differing ancestral histories. Thus eyes have evolved many times along different routes—in worms, snails, insects, and vertebrates. Similarly, wings evolved independently in insects, reptiles, bats, and birds. It seems that nature's creativity is boundless.

Another striking pattern is the repeated occurrence of catastrophes—planetary bifurcation points, perhaps—followed by intense periods of growth and innovation. Thus the disastrous depletion of hydrogen in the Earth's atmosphere over two billion years ago led to one of the greatest evolutionary innovations, the use of water in photosynthesis. Millions of years later this tremendously successful new biotechnology produced a catastrophic pollution crisis by accumulating large amounts of toxic oxygen. The oxygen crisis, in turn, prompted the evolution of oxygen-breathing bacteria, another of life's spectacular innovations. More recently, 245 million years ago the most devastating mass extinctions the world has ever seen were followed rapidly by the evolution of mammals; and 66 million years ago the catastrophe that eliminated the dinosaurs from the face of the Earth cleared the way for the evolution of the first primates and, eventually, the human species.

The Ages of Life

To chart the unfolding of life on Earth, we have to use a geological time scale, on which periods are measured in billions of years. It begins with the formation of the planet Earth, a fireball of molten lava, around 4.5 billion years ago. Geologists and paleontologists have divided those 4.5 billion years into numerous periods and subperiods, labeled by names such as "proterozoic," "paleozoic," "cretaceous," or "pleistocene." Fortunately we do not need to remember any of those technical terms to have an idea of the major stages of life's evolution.

We can distinguish three broad ages in the evolution of life on Earth, each extending for periods between 1 and 2 billion years and each containing several distinct stages of evolution (see table on page 234). The first is the prebiotic age, in which the conditions for the emergence of life were formed. It lasted 1 billion years, from the formation of the Earth to the creation of the first cells, the beginning of life, around 3.5 billion years ago. The second age, extending for a full 2 billion years, is the age of the microcosm, in which bacteria and other microorganisms invented all the basic processes of life and established the global feedback loops for the self-regulation of the Gaia system.

Around 1.5 billion years ago the Earth's modern surface and atmosphere were largely established; microorganisms permeated the air, water, and soil, cycling gases and nutrients through their planetary network, as they do today; and the stage was set for the third age of life, the macrocosm, which saw the evolution of the visible forms of life, including ourselves.

The Origin of Life

During the first billion years after the formation of the Earth, the conditions for the emergence of life gradually fell into place. The primeval fireball was large enough to hold an atmosphere and

Ages of Life	Billion Years Ago	Stages of Evolution
PREBIOTIC AGE	**4.5**	formation of Earth
formation of the		fireball of molten lava
conditions for life		cooling
	4.0	oldest rocks
		condensation of steam
	3.8	shallow oceans
		carbon-based compounds
		catalytic loops, membranes
MICROCOSM	**3.5**	first bacterial cells
evolution of		fermentation
microorganisms		photosynthesis
		sensing devices, motion
		DNA repair
		trading of genes
	2.8	tectonic plates, continents
		oxygen photosynthesis
	2.5	bacteria fully extended
	2.2	first nucleated cells
	2.0	oxygen buildup in atmosphere
	1.8	oxygen breathing
	1.5	Earth surface and atmosphere established
MACROCOSM	1.2	locomotion
evolution of	1.0	sexual reproduction
visible life forms	0.8	mitochondria, chloroplasts
	0.7	early animals
	0.6	shells and skeletons
	0.5	early plants
	0.4	land animals
	0.3	dinosaurs
	0.2	mammals
	0.1	flowering plants
		first primates

contained the basic chemical elements out of which the building blocks of life were to be formed. Its distance from the sun was just right—far enough away for a slow process of cooling and condensation to begin and yet close enough to prevent its gases from being permanently frozen.

After half a billion years of gradual cooling, the steam filling the atmosphere finally condensed; torrential rains fell for thousands of years, and water gathered to form shallow oceans. During this long period of cooling, carbon, the chemical backbone of life, combined rapidly with hydrogen, oxygen, nitrogen, sulfur, and phosphorus to generate an enormous variety of chemical compounds. Those six elements—C, H, O, N, S, P—are now the main chemical ingredients in all living organisms.

For many years scientists debated the likelihood of life emerging from the "chemical soup" that formed as the planet cooled off and the oceans expanded. Several hypotheses of sudden triggering events competed with one another—a dramatic flash of lightning or even a seeding of the Earth with macromolecules by meteorites. Other scientists argued that the odds of any such event having happened are vanishingly small. However, the recent research on self-organizing systems indicates strongly that there is no need to postulate any sudden event.

As Margulis points out, "Chemicals do not combine randomly, but in ordered, patterned ways."[22] The environment on the early Earth favored the formation of complex molecules, some of which became catalysts for a variety of chemical reactions. Gradually different catalytic reactions interlocked to form complex catalytic webs involving closed loops—first cycles, then "hypercycles"—with a strong tendency for self-organization and even self-replication.[23] Once this stage was reached, the direction for prebiotic evolution was set. The catalytic cycles evolved into dissipative structures and, by passing through successive instabilities (bifurcation points), generated chemical systems of increasing richness and diversity.

Eventually these dissipative structures began to form membranes—first, perhaps, from fatty acids without proteins, like the micelles recently produced in the laboratory.[24] Margulis speculates that many different types of membrane-enclosed replicating chemical systems may have arisen, evolved for a while, and then disappeared again before the first cells emerged: "Many dissipative structures, long chains of different chemical reactions, must have evolved, reacted, and broken down before the elegant double helix of our ultimate ancestor formed and replicated with high fidelity."[25] At that moment, about 3.5 billion years ago, the first autopoietic bacterial cells were born, and the evolution of life began.

Weaving the Bacterial Web

The first cells led a precarious existence. The environment around them changed continually, and every hazard presented a new threat to their survival. In the face of all these hostile forces—harsh sunlight, meteorite impacts, volcanic eruptions, droughts, and floods—the bacteria had to trap energy, water, and food to maintain their integrity and stay alive. Each crisis must have wiped out large portions of the first patches of life on the planet and would certainly have extinguished them altogether, had it not been for two vital traits—the abilities of the bacterial DNA to replicate faithfully and to do so with extraordinary speed. Because of their enormous numbers, the bacteria were able, again and again, to respond creatively to all threats and to develop a great variety of adaptive strategies. Thus they gradually expanded, first in the waters and then in the surfaces of sediments and soil.

Perhaps the most important task was to develop a variety of new metabolic pathways for extracting food and energy from the environment. One of the first bacterial inventions was fermentation—the breaking down of sugars and conversion into ATP molecules, the "energy carriers" that fuel all cellular processes.[26] This innovation allowed the fermenting bacteria to live off chemicals in the earth, in mud and water, protected from the harsh sunlight.

Some of the fermenters also developed the ability to absorb nitrogen gas from the air and convert it into various organic compounds. To "fix" nitrogen—in other words, to capture it directly from the air—takes large amounts of energy and is a feat that even today can be performed only by a few special bacteria. Since nitrogen is an ingredient of the proteins in all cells, all living organisms today depend on the nitrogen-fixing bacteria for their survival.

Early on in the age of bacteria, photosynthesis—"undoubtedly the most important single metabolic innovation in the history of life on the planet"[27]—became the primary source of life energy. The first processes of photosynthesis invented by the bacteria were different from those used by plants today. They used hydrogen sulfide, a gas spewed out by volcanoes, instead of water as their source of hydrogen, combined it with sunlight and CO_2 from the air to form organic compounds, and never produced oxygen.

These adaptive strategies not only enabled the bacteria to survive and evolve, but also began to change their environment. In fact, almost from the beginning of their existence, the bacteria established the first feedback loops that would eventually result in the tightly coupled system of life and its environment. Although the chemistry and climate of the early Earth were conducive to life, this favorable state would not have continued indefinitely without bacterial regulation.[28]

As iron and other elements reacted with water, hydrogen gas was released and rose up through the atmosphere, where it broke down into hydrogen atoms. Since these atoms are too light to be held by the Earth's gravity, all the hydrogen would have escaped if this process had continued unchecked, and a billion years later the oceans of the planet would have disappeared. Fortunately life intervened. In the later stages of photosynthesis free oxygen was released into the air, as it is today, and some of it combined with the rising hydrogen gas to form water, thus keeping the planet moist and preventing its oceans from evaporating.

However, the continuing removal of CO_2 from the air in the process of photosynthesis caused another problem. At the beginning of the age of bacteria, the sun was 25 percent less luminous than it is now, and the CO_2 in the atmosphere was very much needed as a greenhouse gas to keep the planetary temperatures in a comfortable range. Had the removal of CO_2 gone on without any compensation, the Earth would have frozen and early bacterial life would have been extinguished.

Such a disastrous course was prevented by the fermenting bacteria, which may have evolved already before the onset of photosynthesis. In the process of producing ATP molecules from sugars, the fermenters also produced methane and CO_2 as waste products. These were emitted into the atmosphere, where they restored the planetary greenhouse. In this way fermentation and photosynthesis became two mutually balancing processes of the early Gaia system.

The sunlight coming through the Earth's early atmosphere still contained burning ultraviolet radiation, but now the bacteria had to balance their protection from exposure with their need for solar energy for photosynthesis. This led to the evolution of numerous sensing systems and of movement. Some bacterial species migrated into waters rich in certain salts that acted as sun filters; others found shelter in sand; yet others developed pigments that absorbed the harmful rays. Many species built huge colonies—multileveled microbial mats in which the top layers got scorched and died but shielded the lower layers with their dead bodies.[29]

In addition to protective filtering the bacteria also developed mechanisms for repairing radiation-damaged DNA, evolving special enzymes for that purpose. Almost all organisms today still possess these repair enzymes—another lasting invention of the microcosmos.[30]

Instead of using their own genetic material for the repair process, bacteria in crowded environments sometimes borrowed DNA fragments from their neighbors. This technique gradually evolved into the constant gene trading that became the most effective avenue of bacterial evolution. In higher forms of life the recombination of genes from different individuals is associated with reproduction, but in the world of bacteria the two phenomena take place independently. Bacterial cells reproduce asexually, but they continually trade genes. In the words of Margulis and Sagan:

> We trade genes "vertically"—through the generations—whereas bacteria trade them "horizontally"—directly to their neighbors in the same generation. The result is that while genetically fluid bac-

teria are functionally immortal, in eukaryotes, sex becomes linked with death.[31]

Because of the small number of permanent genes in a bacterial cell—typically less than 1 percent of those in a nucleated cell—bacteria necessarily work in teams. Different species cooperate and help each other out with complementary genetic material. Large assemblies of such bacterial teams can operate with the coherence of a single organism, performing tasks that none of them can do individually.

By the end of the first billion years after the emergence of life, the Earth was teeming with bacteria. Thousands of biotechnologies had been invented—indeed, most of those known today—and by cooperating and continually trading genetic information the microorganisms had begun to regulate conditions for life on the entire planet, as they still do today. In fact, many of the bacteria living in the early age of the microcosm have survived essentially unchanged to this very day.

During subsequent stages of evolution, the microorganisms formed alliances and coevolved with plants and animals, and today our environment is so interwoven with bacteria that it is almost impossible to say where the inanimate world ends and life begins. We tend to associate bacteria with disease, but they are also vital for our survival, as they are for the survival of all animals and plants. "Beneath our superficial differences we are all of us walking communities of bacteria," write Margulis and Sagan. "The world shimmers, a pointillist landscape made of tiny living beings."[32]

The Oxygen Crisis

As the bacterial web expanded and filled every available space in the waters, rocks, and mud flats of the early planet, its energy needs led to a severe depletion of hydrogen. The carbohydrates that are essential to all life are elaborate structures of carbon, hydrogen, and oxygen atoms. To build these structures the photosynthesizing bacteria took the carbon and oxygen from the air in the form of CO_2, as all plants do today. They also found hydrogen in the air, in the form of hydrogen gas, and in the hydrogen sulfide bubbling up from volcanoes. But the light hydrogen gas kept escaping into space, and eventually the hydrogen sulfide became insufficient.

Hydrogen, of course, exists in great abundance in water (H_2O), but the bonds between hydrogen and oxygen in water molecules are much stronger than those between the two hydrogen atoms in hydrogen gas (H_2) or hydrogen sulfide (H_2S). The photosynthesizing bacteria were not able to break these strong bonds until a special kind of blue-green bacteria invented a new type of photosynthesis that solved the hydrogen problem forever.

The newly evolved bacteria, the ancestors of the modern-day blue-green algae, used sunlight of higher energy (shorter wavelength) to split water molecules into their hydrogen and oxygen components. They took the hydrogen for building sugars and other carbohydrates and emitted the oxygen into the air. This extraction of hydrogen from water, which is one of the planet's most abundant resources, was an extraordinary evolutionary feat with far-reaching implications for the subsequent unfolding of life. Indeed, Lynn Margulis is convinced that "the advent of oxygenic photosynthesis was the singular event that led eventually to our modern environment."[33]

With their unlimited source of hydrogen, the new bacteria were spectacularly successful. They expanded rapidly across the Earth's surface, covering rocks and sand with their blue-green film. Even today they are ubiquitous, growing in ponds and swimming pools, on moist walls and shower curtains—wherever there is sunlight and water.

However, this evolutionary success came at a heavy price. Like all rapidly expanding living systems, the blue-green bacteria produced massive amounts of waste, and in their case this waste was also highly toxic. It was the oxygen gas emitted as a by-product of the new type of water-based photosynthesis. Free oxygen is toxic because it reacts easily with organic matter, producing so-called free radicals that are extremely destructive to carbohydrates and other essential biochemical compounds. Oxygen also reacts easily with atmospheric gases and metals, triggering combustion and corrosion, the two most familiar forms of "oxidizing" (combining with oxygen).

At first the Earth easily absorbed the oxygen waste. There were enough metals and sulfur compounds from volcanic and tectonic sources that quickly captured the free oxygen and prevented it from building up in the air. But after absorbing oxygen for millions of years, the oxidizing metals and minerals became saturated and the toxic gas began to accumulate in the atmosphere.

About two billion years ago the oxygen pollution resulted in a catastrophe of unprecedented global proportions. Numerous species were wiped out completely, and the entire bacterial web had to fundamentally reorganize itself to survive. Many protective devices and adaptive strategies evolved, and finally the oxygen crisis led to one of the greatest and most successful innovations in the entire history of life:

> In one of the greatest coups of all time, the [blue-green] bacteria invented a metabolic system that *required* the very substance that had been a deadly poison. . . . The breathing of oxygen is an ingeniously efficient way of channeling and exploiting the reactivity of oxygen. It is essentially controlled combustion that breaks down organic molecules and yields carbon dioxide, water, and a great deal of energy in the bargain. . . . The microcosm did more than adapt: it evolved an oxygen-using dynamo that changed life and its terrestrial dwelling place forever.[34]

With this spectacular invention the blue-green bacteria had two complementary mechanisms at their disposal—the generation of free oxygen through photosynthesis and its absorption through respiration—and thus they could begin to set up the feedback loops that would henceforth regulate the atmosphere's oxygen content, maintaining it at the delicate balance that enabled new oxygen-breathing forms of life to evolve.[35]

The proportion of free oxygen in the atmosphere eventually stabilized at 21 percent, a value determined by its range of flammability. If it dropped to below 15 percent, *nothing* would burn. Organisms could not breathe and would asphyxiate. If the oxygen in the air rose to above 25 percent, *everything* would burn. Combustion would occur spontaneously and fires would rage around the planet. Accordingly, Gaia has kept the atmospheric oxygen at the level most comfortable for all plants and animals for millions of years. In addition, a layer of ozone (three-atom oxygen molecules) gradually built up at the top of the atmosphere and from then on protected life on Earth from the sun's harsh ultraviolet rays. Now the stage was set for the evolution of the larger forms of life—fungi, plants, and animals—which occurred in relatively short periods of time.

The Nucleated Cell

The first step toward higher forms of life was the emergence of symbiosis as a new avenue for evolutionary creativity. This occurred around 2.2 billion years ago and led to the evolution of eukaryotic ("nucleated") cells, which became the fundamental components of all plants and animals. Nucleated cells are much larger and far more complex than bacteria. Whereas the bacterial cell contains a single loose strand of DNA floating freely in the cell fluid, the DNA in a eukaryotic cell is coiled tightly into chromosomes, which are confined by a membrane inside the cell nucleus. The amount of DNA in nucleated cells is several hundred times that found in bacteria.

The other striking characteristic of the nucleated cell is an abundance of organelles—oxygen-using smaller cell parts that carry out a variety of highly specialized functions.[36] The sudden appearance of nucleated cells in the history of evolution and the discovery that their organelles are distinct self-reproducing organisms led Lynn Margulis to the conclusion that nucleated cells have evolved through long-term symbiosis, the permanent living together of various bacteria and other microorganisms.[37]

The ancestors of the mitochondria and other organelles may have been vicious bacteria that invaded larger cells and reproduced inside them. Many of the invaded cells would have died, taking the invaders with them. However, some of the predators did not kill their hosts outright but began to cooperate with them, and eventually natural selection allowed only the cooperators to survive and evolve further. Nuclear membranes may have evolved to protect the host cells' genetic material from attack by the invaders.

Over millions of years the cooperative relationships became ever more coordinated and interwoven, organelles reproducing offspring well adapted to living within larger cells and larger cells becoming ever more dependent on their lodgers. Over time these bacterial communities became so utterly interdependent that they functioned as single integrated organisms:

> Life had moved another step, beyond the networking of free genetic transfer to the synergy of symbiosis. Separate organisms blended together, creating new wholes that were greater than the sum of their parts.[38]

The recognition of symbiosis as a major evolutionary force has profound philosophical implications. All larger organisms, including ourselves, are living testimonies to the fact that destructive practices do not work in the long run. In the end the aggressors always destroy themselves, making way for others who know how to cooperate and get along. Life is much less a competitive struggle for survival than a triumph of cooperation and creativity. Indeed, since the creation of the first nucleated cells, evolution has proceeded through ever more intricate arrangements of cooperation and coevolution.

The avenue of evolution through symbiosis allowed the new forms of life to use well-tested specialized biotechnologies over and over again in different combinations. For example, whereas bacteria obtain their food and energy by a great variety of ingenious methods, only one of their numerous metabolic inventions is used by animals—that of oxygen breathing, the specialty of the mitochondria.

Mitochondria are also present in plant cells, which in addition contain the so-called chloroplasts, the green "solar stations" responsible for photosynthesis.[39] These organelles are remarkably similar to the blue-green bacteria, the inventors of oxygen photosynthesis, who in all likelihood were their ancestors. Margulis speculates that those all-pervasive bacteria were routinely eaten by other microorganisms and that some variations must have resisted being digested by their hosts.[40] Instead they adapted to their new environments while continuing to produce energy through photosynthesis, upon which the larger cells soon became dependent.

While their new symbiotic relationships gave the nucleated cells access to the efficient use of sunlight and oxygen, they also gave them a third great evolutionary advantage—the capability of movement. Whereas the components of a bacterial cell float around slowly and passively in the cell fluid, those in a nucleated cell seem to move decisively; the cell fluid streams along, and the entire cell may expand and contract rhythmically or move rapidly as a whole, as, for example, in the case of blood cells.

Like so many other life processes, rapid motion was invented by bacteria. The fastest member of the microcosm is a tiny, hairlike creature called *spirochete* ("coiled hair"), also known as the "corkscrew bacterium," which spirals in rapid motion. By attaching themselves symbiotically to larger cells, the rapidly moving corkscrew bacteria gave those cells the tremendous advantages of locomotion—the ability to avoid danger and seek out food. Over time the corkscrew bacteria progressively lost their distinct traits and evolved into the well-known "cell whips"—*flagellae, cilia,* and the like—that propel a wide variety of nucleated cells with undulating or whipping motions.

The combined advantages of the three types of symbiosis described in the preceding paragraphs created a burst of evolutionary activity that generated a tremendous diversity of eukaryotic cells. With their two effective means of energy production and their dramatically increased mobility, the new symbiotic life forms migrated to many new environments, evolving into the primeval plants and animals that would eventually leave the water and take over the land.

As a scientific hypothesis the concept of symbiogenesis—the creation of new forms of life through the merging of different species—is barely thirty years old. But as a cultural myth the idea seems to be as old as humanity itself.[41] Religious epics, legends, fairy tales, and other mythical stories around the world are full of fantastic creatures—sphinxes, mermaids, griffons, centaurs, and more—born from the blending of two or more species. Like the new eukaryotic cells, these creatures are made of components that are entirely familiar, but their combinations are novel and startling.

Depictions of these hybrid beings are often frightening, but many of them, curiously, are seen as bearers of good fortune. For example, the god Ganesha, who has a human body with an elephant head, is one of the most revered deities in India, worshiped as a symbol of good luck and a helper in overcoming obstacles. Somehow the collective human unconscious seems to have known from ancient times that long-term symbioses are profoundly beneficial for all life.

Evolution of Plants and Animals

The evolution of plants and animals out of the microcosm proceeded through a succession of symbioses, in which the bacterial inventions from the previous two billion years were combined in endless expressions of creativity until viable forms were selected to survive. This evolutionary process is characterized by increasing specialization—from the organelles in the first eukaryotes to the highly specialized cells in animals.

An important aspect of cell specialization is the invention of sexual reproduction, which occurred about one billion years ago. We tend to think of sex and reproduction as being closely associated, but Margulis points out that the complex dance of sexual reproduction consists of several distinct components that evolved independently and only gradually became interlinked and unified.[42]

The first component is a type of cell division, called *meiosis* ("diminution"), in which the number of chromosomes in the nucleus is reduced by exactly half. This creates specialized egg and sperm cells. These cells are then fused in the act of fertilization, in which the normal number of chromosomes is restored and a new cell, the fertilized egg, is created. This cell then divides repeatedly in the growth and development of a multicellular organism.

The fusion of genetic material from two different cells is widespread among bacteria, where it takes place as a continual trading of genes that is not linked to reproduction. In the early plants and animals reproduction and the fusion of genes became linked and subsequently evolved into elaborate processes and rituals of fertilization. Gender was a later refinement. The first germ cells—sperm and egg—were almost identical, but over time they evolved into small fast-moving sperm cells and large stationary eggs. The connection of fertilization with the formation of embryos came even later in the evolution of animals. In the world of plants fertilization led to intricate patterns of coevolution of flowers, insects, and birds.

As the specialization of cells continued in larger and more complex forms of life, the capability of self-repair and regeneration diminished progressively. Flatworms, polyps, and starfish can regenerate almost their entire bodies from small fractions; lizards, salamanders, crabs, lobsters, and many insects are still able to grow back lost organs or limbs; but in higher animals regeneration is limited to renewing tissues in the healing of injuries. As a consequence of this loss of regenerative capabilities, all large organisms age and eventually die. However, with sexual reproduction life has invented a new type of regenerative process, in which entire organisms are formed anew again and again, returning in every "generation" to a single nucleated cell.

Plants and animals are not the only multicellular creatures in the living world. Like other traits of living organisms, multicellularity evolved many times in many lineages of life, and today there still exist several kinds of multicellular bacteria and many multicellular protists (microorganisms with nucleated cells). Like animals and plants, most of these multicellular organisms are formed by successive cell divisions, but some may be generated by an aggregation of cells from different sources but of the same species.

A spectacular example of such aggregations is the slime mold, an organism that is macroscopic but is technically a protist. A slime mold has a complex life cycle involving a mobile (animal-like) and an immobile (plant-like) phase. In the animal-like phase it starts out as a multitude of single cells, commonly found in forests under rotting logs and damp leaves, where they feed on other microorganisms and decaying vegetation. The cells often eat so much and divide so rapidly that they deplete the entire food supply in their environment. When this happens they aggregate into a cohesive mass of thousands of cells, resembling a slug and capable of creeping across the forest floor in amoebalike movements. When it has found a new source of food, the mold enters its plantlike phase, developing a stalk with a fruiting body and looking very much like a fungus. Finally the fruit capsule bursts, shooting out thousands of dry spores from which new individual cells are born, to move about independently in the search for food, starting a new cycle of life.

Among the many multicellular organizations that evolved out of tightly knit communities of microorganisms, three—plants, fungi, and animals—have been so successful in reproducing, diversifying, and expanding over the Earth that they are classified by biologists as "kingdoms," the broadest category of living organisms. All in all there are five of these kingdoms—bacteria (microorganisms without cell nuclei), protists (microorganisms with nucleated cells), plants, fungi, and animals.[43] Each of the kingdoms is divided into a hierarchy of subcategories, or *taxa,* beginning with *phylum* and ending with *genus* and *species.*

The theory of symbiogenesis has allowed Lynn Margulis and her colleagues to base the classification of living organisms on clear evolutionary relationships. Figure 10-1 shows in simplified form how the protists, plants, fungi, and animals all evolved from the bacteria through a series of successive symbioses, described in more detail in the following pages.

When we follow the evolution of plants and animals we find ourselves in the macrocosm and have to shift our time scale from billions of years to millions. The earliest animals evolved around 700 million years ago, and the earliest plants emerged about 200 million years later. Both evolved first in water and came ashore 400–450 million years ago, the plants preceding the animals on land by several million years. Plants and animals both developed huge multicellular organisms, but whereas intercellular communication is minimal in plants, animal cells are highly specialized and

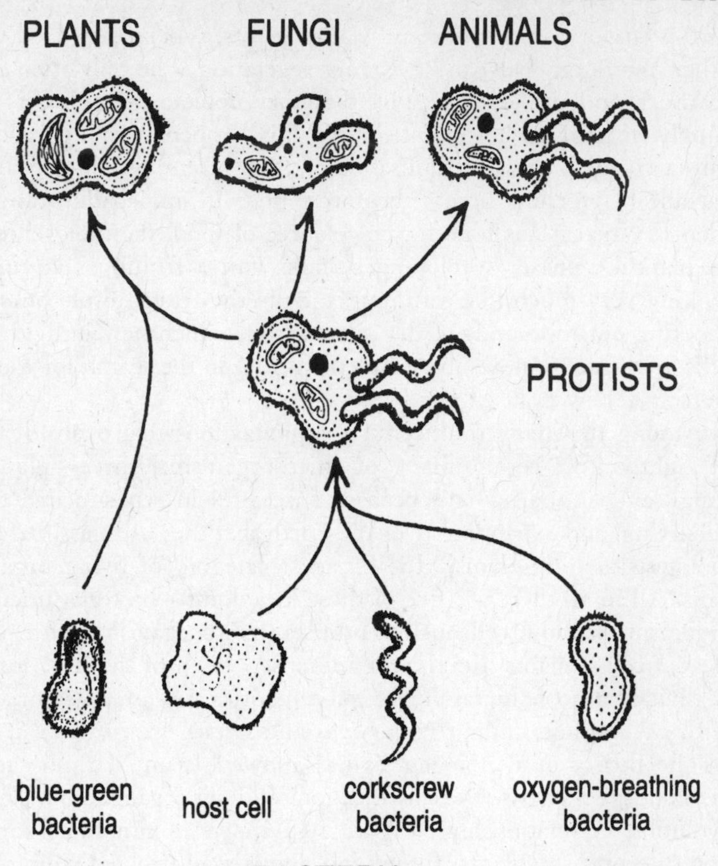

Figure 10-1
Evolutionary relationships among the five kingdoms of life.

tightly interconnected by a variety of elaborate links. Their mutual coordination and control was greatly increased by the very early creation of nervous systems, and by 620 million years ago tiny animal brains had evolved.

The ancestors of plants were thready masses of algae that dwelled in sunlit shallow waters. Occasionally their habitats would dry up, and eventually some algae managed to survive, reproduced, and turned into plants. Those early plants, rather like today's mosses, had neither stems nor leaves. To survive on land it was crucial for them to develop sturdy structures so that they would not collapse and dry out. They did so by creating lignin, a material for cell walls that enabled plants to grow sturdy stems and branches, as well as vascular systems to draw water up from the roots.

The major challenge of the new environment on land was the shortage of water. The creative answer of plants was to enclose their embryos in protective, drought-resistant seeds, so that they could wait with their development until they found themselves in an appropriately moist environment. For over 100 million years, while the first land animals, the amphibians, evolved into reptiles and dinosaurs, lush tropical forests of "seed ferns"—seed-bearing trees that resembled giant ferns—covered large portions of the Earth.

About 200 million years ago glaciers appeared on several continents, and the seed ferns could not survive the long, cold winters. They were replaced by evergreen conifers, similar to our present-day fir and spruce, whose greater resistance to cold allowed them to survive the winters and even to expand into higher alpine regions. One hundred million years later flowering plants whose seeds were enclosed in fruits began to appear.

From the beginning these new flowering plants coevolved with animals, who enjoyed eating their nutritious fruits and in exchange disseminated the undigested plant seeds. These cooperative arrangements have continued to develop and now also include human growers who not only distribute plant seeds, but also clone seedless plants for their fruits. As Margulis and Sagan observe, "Plants indeed seem very adept at seducing us animals, having tricked us into doing for them one of the few things we can do that they cannot: move."[44]

Conquering the Land

The first animals evolved in water from globular and wormlike masses of cells. They were still very small, but some of them formed communities that collectively built huge coral reefs with their calcium deposits. Lacking any hard parts or internal skeletons, the early animals completely disintegrated at death, but a hundred million years later their descendants produced a wealth of exquisite shells and skeletons that left clear imprints in well-preserved fossils.

For animals, the adaptation to life on land was an evolutionary feat of staggering proportions, requiring drastic changes in all organ systems. The greatest problem in the absence of water, of course, was desiccation; but there were a host of other problems as well. There was enormously more oxygen in the atmosphere than in the oceans, which required different organs for breathing; different types of skin were necessary for protection against unfiltered sunlight; and stronger muscles and bones were needed to deal with gravity in the absence of buoyancy.

To ease the transition to these totally different surroundings, animals invented a most ingenious trick. They took their former environment with them for their young. To this day the animal womb simulates the wetness, buoyancy, and salinity of the ancient marine environment. Moreover, the salt concentrations in the mammal blood and other bodily fluids are remarkably similar to those in the oceans. We came out of the ocean more than 400 million years ago, but we never completely left the seawater behind. We still find it in our blood, sweat, and tears.

Another major innovation that became vital for living on land had to do with the regulation of calcium. Calcium plays a central role in the metabolism of all nucleated cells. In particular it is crucial to the operation of muscles. For these metabolic processes to work, the amount of calcium must be kept at precise levels, which are much lower than the calcium levels in seawater. Therefore marine animals from the very beginning had to continually remove all excess calcium. The early smaller animals simply excreted their calcium waste, sometimes piling it up in enormous coral reefs. As larger animals evolved, they began to stockpile the excess calcium around and inside themselves, and these deposits eventually turned into shells and skeletons.

As the blue-green bacteria had transformed a toxic pollutant, oxygen, into a vital ingredient for their further evolution, so the early animals transformed another major pollutant, calcium, into building materials for new structures that gave them tremendous selective advantages. Shells and other hard parts were used to fend off predators, while skeletons emerged first in fish and subsequently evolved into the essential support structures of all large animals.

Around 580 million years ago, at the beginning of the so-called Cambrian period, there was such a profusion of fossils with beautiful clear imprints of shells, rigid coats, and skeletons that paleontologists believed for a long time that these Cambrian fossils marked the beginning of life. Sometimes they were even viewed as records of God's first acts of creation. It is only within the last three decades that the traces of the microcosm have been revealed in so-called chemical fossils.[45] These show conclusively that the origins of life predate the Cambrian period by almost three billion years.

Evolutionary experiments with calcium deposits led to a great diversity of forms—tubular "sea squirts" with spinal columns but no bones, fishlike creatures with external armors but without jaws, lungfish that breathed both water and air, and many more. The first vertebrate creatures with backbones and a braincase shielding the nervous system probably evolved around 500 million years ago. Among them was a lineage of fish with lungs, stubby fins, jaws, and a frog-like head, which crawled along the shores and eventually evolved into the first amphibians. The amphibians—frogs, toads, salamanders, and newts—are the evolutionary link between water and land animals. They are the first terrestrial vertebrates, but even today they begin their life cycle as water-breathing tadpoles.

The first insects came ashore around the same time as the amphibians and may even have encouraged some fish to feed on them and follow them out of the water. On land the insects exploded into an enormous variety of species. Their small size and high reproductive rates allowed them to adapt to almost any environment by developing a fabulous diversity of body structures and

Evolution of Plants and Animals

Million Years Ago	Stages of Evolution
700	early animals
620	first animal brains
580	shells and skeletons
500	vertebrates
450	plants come ashore
400	amphibians and insects come ashore
350	seed ferns
300	fungi
250	reptiles
225	conifers, dinosaurs
200	mammals
150	birds
125	flowering plants
70	extinction of dinosaurs
65	early primates
35	monkeys
20	apes
10	great apes
4	upright walking "Southern apes"

ways of life. There are about 750,000 known species of insects today, three times as many as all other animal species together.

During the 150 million years after they left the sea, the amphibians evolved into reptiles, endowed with several strong selective advantages—powerful jaws, drought-resistant skin, and most important, a new kind of eggs. As the mammals would in their wombs later on, the reptiles encapsulated the former marine environment in large eggs, in which their offspring could prepare themselves fully for spending their entire life cycles on land. With these innovations reptiles rapidly conquered the land and evolved into numerous varieties. The many types of lizards that still exist today, including the limbless snakes, are descendants of those ancient reptiles.

While the first lineage of fish crawled out of the water and turned into amphibians, shrubs and trees were already thriving on land, and when the amphibians evolved into reptiles they lived in lush tropical forests. At the same time, a third type of multicellular organism, the fungi, had come ashore. Fungi are plantlike and yet so different from plants that they are classified as a separate kingdom, which displays a variety of fascinating properties.[46] They lack the green chlorophyll for photosynthesis and do not eat and digest, but absorb their nutrients directly as chemicals. Unlike plants, fungi do not have vascular systems for forming roots, stems, and leaves. They have very distinctive cells, which may contain several nuclei and are separated by thin walls through which the cell fluid can flow easily.

Fungi emerged more than 300 million years ago and expanded in close coevolution with plants. Virtually all plants that grow in the soil rely on a tiny fungus in their roots for the absorption of nitrogen. In a forest the roots of all the trees are interconnected by an extensive fungal network, which occasionally comes up through the earth as mushrooms. Without fungi the primeval tropical forests could not have existed.

Thirty million years after the appearance of the first reptiles, one of their lineages evolved into dinosaurs (a Greek term meaning "terrible lizards"), which seem to hold endless fascination for humans of all ages. They came in a great variety of sizes and shapes. Some had body armors with horny beaks, like modern turtles, or horns. Some were herbivores, others were carnivores. Like the other reptiles, dinosaurs were egg-laying animals. Many built nests, and some even developed wings and eventually, around 150 million years ago, evolved into birds.

At the time of the dinosaurs the expansion of reptiles was in full swing. The land and waters were populated by snakes, lizards, and sea turtles, as well as by sea serpents and several species of dinosaurs. Around 70 million years ago the dinosaurs and many other species suddenly disappeared, most likely because of the impact of a giant meteorite measuring seven miles across. The catastrophic explosion generated an enormous cloud of dust, blocking out sunlight for a prolonged period and drastically changing worldwide weather patterns, which the huge dinosaurs could not survive.

Caring for the Young

About 200 million years ago a warm-blooded vertebrate evolved from the reptiles and diversified into a new class of animals that would eventually bring forth our ancestors, the primates. The females of these warm-blooded animals no longer enclosed their embryos in eggs but instead nourished them inside their own bodies. After birth the young were relatively helpless and were nursed by their mothers. Because of this distinctive behavior, which includes nursing with milk secreted from mammary glands, this class of animals is known as "mammals." Around 50 million years later another lineage of warm-blooded vertebrates, the newly evolved birds, also began to feed and teach their vulnerable offspring.

The first mammals were small nocturnal creatures. Whereas the reptiles, unable to regulate their body temperatures, were sluggish during the cool nights, the mammals evolved the ability to maintain their body warmth at relatively constant levels independent of their surroundings and thus remained alert and active at night. They also transformed some of their skin cells into hair, which insulated them further and allowed them to migrate from the tropics to colder climates.

The early primates, known as prosimians ("premonkeys"), evolved in the tropics around 65 million years ago from nocturnal, insect-eating mammals that lived in trees and looked somewhat like squirrels. Today's prosimians are small forest animals, mostly nocturnal and still living in trees. To jump from branch to branch at night, those early insect-eating tree dwellers developed keen eyesight, and in some species the eyes shifted gradually to a frontal position, which was crucial to developing three-dimensional vision—a decisive advantage for judging distances in trees. Other well-known primate characteristics that evolved from their tree-climbing skills are clinging hands and feet, flat fingernails, opposable thumbs, and big toes.

Unlike other animals, the prosimians were not anatomically specialized and therefore were always threatened by enemies. However, they made up for their lack of specialization by developing greater dexterity and intelligence. Their fear of enemies, constant running and hiding, and active night life encouraged cooperation and led to the social behavior that is characteristic of all higher primates. In addition, the habit of protecting themselves by making frequent loud noises gradually evolved into vocal communication.

Most primates are insect eaters or vegetarians, feeding on nuts, fruits, and grasses. At times, when not enough nuts and fruits were available in the trees, the early primates would have left the protective branches and come down to the ground. Looking anxiously for enemies over tall grasses, they would assume an upright posture for brief moments before returning to a crouched position, as baboons still do today. This ability to stand upright, even for short moments, represented a strong selective advantage, as it allowed the primates to use their hands for gathering food, wielding sticks, or throwing rocks to defend themselves. Gradually their feet became flatter, their manual dexterity increased, and the use of primitive tools and weapons stimulated brain growth, and thus some of the prosimians evolved into monkeys and apes.

The evolutionary line of the monkeys diverged from that of the prosimians around 35 million years ago. Monkeys are diurnal animals, generally with flatter and more expressive faces than those of prosimians, and they usually walk or run on four legs. Around 20 million years ago the line of the apes split from that of the monkeys, and after another 10 million years our immediate ancestors, the great apes—orangutans, gorillas, and chimpanzees—came into their own.

All apes are forest dwellers, and most of them spend at least some of the time in trees. Gorillas and chimpanzees are the most terrestrial of the apes, traveling on all fours by "knuckle walking"—that is, leaning on the knuckles of their forelimbs. Most apes are also able to walk on two legs for short distances. Like humans, apes have broad, flat chests, and arms capable of reaching up and backward from the shoulder. This enables them to move in trees by swinging from branch to branch arm over arm, a feat of which monkeys are not capable. The brains of the great apes are much more complex than those of monkeys, and thus their intelligence is far superior. The ability to use and, to a limited extent, even make tools is characteristic of the great apes.

Around 4 million years ago a chimpanzee species in the African tropics evolved into an upright walking ape. This primate species, which became extinct a million years later, was quite similar to the other great apes, but because of its upright gait it has been classified as a "hominid," which, according to Lynn Margulis, is unjustified on purely biological grounds:

> Objective scholars, if they were whales or dolphins, would place humans, chimpanzees, and orangutans in the same taxonomic group. There is no physiological basis for the classification of human beings into their own family. . . . Human beings and chimps are far more alike than any two arbitrarily chosen genera of beetles. Nonetheless, animals that walk upright with their hands dangling free are aggrandizingly defined as hominids . . . not apes.[47]

The Human Adventure

Having followed the unfolding of life on Earth from its very beginnings, we cannot help feeling a special sense of excitement when we arrive at the stage where the first apes stand up and walk on two legs, even though this may not be justified scientifically. As we learn how reptiles evolved into warm-blooded vertebrates who care for their young; how the first primates developed flat fingernails, opposable thumbs, and the beginnings of vocal communication; and how the apes developed humanlike chests and arms, complex brains, and tool-making capabilities, we can trace the gradual emergence of our human characteristics. And when we reach the stage of upright walking apes with free hands, we feel that now the human evolutionary adventure begins in earnest. To follow it closely, we must shift our time scale once more, this time from millions of years to thousands.

The upright walking apes, which became extinct around 1.4 million years ago, all belonged to the genus *Australopithecus*. The name, derived from the Latin *australis* ("southern") and the Greek *pithekos* ("ape"), means "Southern ape" and is a tribute to the first discoveries of fossils belonging to this genus in South Africa. The oldest species of these Southern apes is known as *Australopithecus afarensis,* named after fossil finds in the Afar region in Ethiopia that included the famous skeleton called "Lucy." They were lightly built primates, perhaps 4.5 feet tall, and probably as intelligent as present-day chimpanzees.

After almost 1 million years of genetic stability, from around 4 to around 3 million years ago, the first species of Southern apes evolved into several more heavily built species. These included two early human species that coexisted with the Southern apes in Africa for several hundred thousand years, until the latter became extinct.

An important difference between human beings and the other primates is that human infants need much longer to pass into childhood, and human children longer again to reach puberty and adulthood, than any of the apes. Whereas the young of other

Human Evolution

Years Ago	Stages of Evolution
4 million	*Australopithecus afarensis*
3.2 million	"Lucy" (*Australopithecus afarensis*)
2.5 million	several *Australopithecus* species
2 million	*Homo habilis*
1.6 million	*Homo erectus*
1.4 million	Australopithecines become extinct
1 million	*Homo erectus* settles in Asia
400,000	*Homo erectus* settles in Europe
	Homo sapiens begins to evolve
250,000	archaic forms of *Homo sapiens*
	Homo erectus becomes extinct
125,000	*Homo neanderthalensis*
100,000	*Homo sapiens* fully evolved in Africa and Asia
40,000	*Homo sapiens* (Cro-Magnon) fully evolved in Europe
35,000	Neanderthals become extinct; *Homo sapiens* remains the single surviving human species

mammals develop fully in the womb and leave it ready for the outside world, our infants are incompletely formed at birth and utterly helpless. Compared with other animals, human infants seem to be born prematurely.

This observation is the basis of the widely accepted hypothesis that the premature births of some apes may have been decisive in triggering human evolution.[48] Because of genetic changes in the timing of development, the prematurely born apes may have retained their youthful traits longer than others. Ape couples with those characteristics, known as *neoteny* ("extension of the new"), would have given birth to more prematurely born children, who would have retained even more youthful traits. Thus an evolutionary trend may have been started that eventually resulted in a relatively hairless species whose adults in many ways resemble the embryos of apes.

According to this hypothesis, the helplessness of the prematurely born infants played a crucial role in the transition from apes to humans. These newborns required supportive families, which may have formed the communities, nomadic tribes, and villages that became the foundations of human civilization. Females selected males who would take care of them while they nursed and protected their infants. Eventually the females no longer went into heat at specific times, and since they could now be sexually receptive at any time, the males caring for their families may have changed their sexual habits as well, decreasing their promiscuity in favor of new social arrangements.

At the same time, the freedom of the hands to make tools, wield weapons, and throw rocks stimulated the continuing brain growth that is characteristic of human evolution and may even have contributed to the development of language. As Margulis and Sagan describe it:

> By throwing rocks, and stunning or killing small prey, early humans were catapulted into a new evolutionary niche. The skills necessary to plot the trajectories of projectiles, to kill at a distance, were dependent on an increase in the size of the left hemisphere of the brain. Language abilities (which have been associated with the left side of the brain . . .) may have fortuitously accompanied such an increase in brain size.[49]

The first human descendants of the Southern apes emerged in East Africa around 2 million years ago. They were a small slender species with markedly expanded brains, which enabled them to develop tool-making skills far superior to those of any of their ape ancestors. This first human species was therefore given the name *Homo habilis* ("skillful human"). By 1.6 million years ago *Homo habilis* had evolved into a more robust and larger species, whose brain had expanded further. Known as *Homo erectus* ("upright human"), this species persisted well over a million years and became far more versatile than its predecessors, adapting its technologies and ways of life to a wide range of environmental conditions. There are indications that these early humans may have gained control of fire around 1.4 million years ago.

Homo erectus was the first species to leave the comfortable African tropics and migrate into Asia, Indonesia, and Europe, settling in Asia around 1 million years ago and in Europe around 400,000 years ago. Far away from their African homeland, the early humans had to endure extremely harsh climatic conditions that had a strong impact on their further evolution. The entire evolutionary history of the human species, from the emergence of *Homo habilis* to the agricultural revolution almost 2 million years later, coincided with the famous ice ages.

During the coldest periods sheets of ice covered large parts of Europe and the Americas, as well as small areas in Asia. These extreme glaciations were interrupted repeatedly by periods during which the ice retreated and gave way to relatively mild climates. However, large-scale floods, caused by the melting of the ice caps during the interglacial periods, were additional threats to animals and humans alike. Many animal species of tropical origin became extinct and were replaced by more robust, woolly species—oxen, mammoths, bison, and the like—which could withstand the harsh conditions of the ice ages.

The early humans hunted those animals with stone axes and spearheads, feasted on them by the fire in their caves, and used the animals' furs to protect themselves from the bitter cold. Hunting together, they also shared their food, and this sharing of food became another catalyst for human civilization and culture, eventually bringing forth the mythical, spiritual, and artistic dimensions of human consciousness.

Between 400,000 and 250,000 years ago *Homo erectus* began to evolve into *Homo sapiens* ("wise human"), the species to which we modern humans belong. This evolution occurred gradually and included several transitional species, which are referred to as archaic *Homo sapiens.* By 250,000 years ago *Homo erectus* was extinct; the transition to *Homo sapiens* was complete around 100,000 years ago in Africa and Asia and around 35,000 years ago in Europe. From that time on, fully modern humans have remained as the single surviving human species.

While *Homo erectus* gradually evolved into *Homo sapiens,* a different line branched off in Europe and evolved into the classic

Neanderthal form around 125,000 years ago. Named after the Neander Valley in Germany, where the first specimen was found, this distinct species persisted until 35,000 years ago. The unique anatomical features of the Neanderthals—they were stocky and robust, with massive bones, low sloping foreheads, heavy jaws, and long, protruding front teeth—were probably due to the fact that they were the first humans to spend long periods in extremely cold environments, having emerged at the onset of the most recent ice age. The Neanderthals settled in southern Europe and Asia, where they left behind signs of ritualized burials in caves decorated with a variety of symbols and of cults involving the animals they hunted. By 35,000 years ago they had either become extinct or had merged with the evolving species of modern humans.

The human evolutionary adventure is the most recent phase in the unfolding of life on Earth, and for us, naturally, it holds a special fascination. However, from the perspective of Gaia, the living planet as a whole, the evolution of human beings has been a very brief episode so far and may even come to an abrupt end in the near future. To demonstrate how late the human species arrived on the planet, the Californian environmentalist David Brower has devised an ingenious narrative by compressing the age of the Earth into the six days of the biblical creation story.[50]

In Brower's scenario the Earth is created on Sunday at midnight. Life in the form of the first bacterial cells appears on Tuesday morning around 8:00 A.M. For the next two and a half days the microcosm evolves, and by Thursday at midnight it is fully established, regulating the entire planetary system. On Friday around 4:00 P.M., the microorganisms invent sexual reproduction, and on Saturday, the last day of creation, all the visible forms of life evolve.

Around 1:30 A.M. on Saturday the first marine animals are formed, and by 9:30 A.M. the first plants come ashore, followed two hours later by amphibians and insects. At ten minutes before five in the afternoon, the great reptiles appear, roam the Earth in lush tropical forests for five hours, and then suddenly die out around 9:45 P.M. In the meantime the mammals have arrived on the Earth in the late afternoon, around 5:30, and the birds in the evening, around 7:15.

Shortly before 10:00 P.M. some tree-dwelling mammals in the tropics evolve into the first primates; an hour later some of those evolve into monkeys; and around 11:40 P.M. the great apes appear. Eight minutes before midnight the first Southern apes stand up and walk on two legs. Five minutes later they disappear again. The first human species, *Homo habilis,* appears four minutes before midnight, evolves into *Homo erectus* half a minute later, and into the archaic forms of *Homo sapiens* thirty seconds before midnight. The Neanderthals command Europe and Asia from fifteen to four seconds before midnight. The modern human species, finally, appears in Africa and Asia eleven seconds before midnight and in Europe five seconds before midnight. Written human history begins around two-thirds of a second before midnight.

By 35,000 years ago the modern species of *Homo sapiens* had replaced the Neanderthals in Europe and evolved into a subspecies known as Cro-Magnon—named after a cave in southern France—to which all modern humans belong. The Cro-Magnons were anatomically identical to us, had fully developed language, and brought forth a veritable explosion of technological innovations and artistic activities. Finely crafted tools of stone and bone, jewelry of shell and ivory, and magnificent paintings on the walls of damp, inaccessible caves are vivid testimonies to the cultural sophistication of those early members of the modern human race.

Until recently archaeologists believed that the Cro-Magnons developed their cave art gradually, beginning with rather crude and clumsy drawings and reaching their height with the famous paintings at Lascaux around 16,000 years ago. However, the sensational discovery of the Chauvet cave in December 1994 forced scientists to radically revise their ideas. This large cave in the Ardèche region of southern France consists of a maze of underground chambers filled with over three hundred highly accomplished paintings. The style is similar to the art at Lascaux, but careful radiocarbon dating has shown that the paintings at Chauvet are at least 30,000 years old.[51]

The figures, painted in ocher, hues of charcoal, and red hematite, are symbolic and mythological images of lions, mammoths, and other dangerous animals, many of them leaping or running across large panels. Specialists in ancient rock art have been amazed by the sophisticated techniques—shading, special angles, staggering of figures, and so on—used by the cave artists to portray motion and perspective. In addition to the paintings, the Chauvet cave also contained a wealth of stone tools and ritualistic objects, including an altarlike stone slab with a bear skull placed on it. Perhaps the most intriguing find is a black drawing of a shamanistic creature, half human and half bison, in the innermost, darkest part of the cave.

The unexpectedly early date of those magnificent paintings means that high art was an integral part of the evolution of modern humans from the very beginning. As Margulis and Sagan point out:

> Such paintings alone clearly mark the presence of modern *Homo sapiens* on earth. Only people paint, only people plan expeditions to the rear ends of damp, dark caves in ceremony. Only people bury their dead with pomp. The search for the historical ancestor of man is the search for the story-teller and the artist.[52]

This means that a proper understanding of human evolution is impossible without understanding the evolution of language, art, and culture. In other words, we must now turn our attention to mind and consciousness, the third conceptual dimension of the systems view of life.

11

Bringing Forth a World

In the emerging theory of living systems mind is not a thing, but a process. It is cognition, the process of knowing, and it is identified with the process of life itself. This is the essence of the Santiago theory of cognition, proposed by Humberto Maturana and Francisco Varela.[1]

The identification of mind, or cognition, with the process of life is a radically new idea in science, but it is also one of the deepest and most archaic intuitions of humanity. In ancient times the rational human mind was seen as merely one aspect of the immaterial soul, or spirit. The basic distinction was not between body and mind, but between body and soul, or body and spirit. While the differentiation between soul and spirit was fluid and fluctuated over time, both originally unified in themselves two concepts—that of the force of life and that of the activity of consciousness.[2]

In the languages of ancient times both of these ideas are expressed through the metaphor of the breath of life. Indeed, the etymological roots of "soul" and "spirit" mean "breath" in many antique languages. The words for "soul" in Sanskrit *(atman),* Greek *(pneuma),* and Latin *(anima)* all mean "breath." The same is true of the word for "spirit" in Latin *(spiritus),* in Greek *(psyche),* and in Hebrew *(ruah).* These, too, mean "breath."

The common ancient intuition behind all these words is that of soul or spirit as the breath of life. Similarly, the concept of cognition in the Santiago theory goes far beyond the rational mind, as it includes the entire process of life. Describing it as the breath of life is a perfect metaphor.

Cognitive Science

Like the concept of "mental process" formulated independently by Gregory Bateson,[3] the Santiago theory of cognition has its roots in cybernetics. It was developed within an intellectual movement that approaches the scientific study of mind and knowledge from a systemic, interdisciplinary perspective beyond the traditional frameworks of psychology and epistemology. This new approach, which has not yet crystallized into a mature scientific field, is increasingly referred to as "cognitive science."[4]

Cybernetics provided cognitive science with the first model of cognition. Its premise was that human intelligence resembles computer "intelligence" to such an extent that cognition can be defined as information processing—that is, as the manipulation of symbols based on a set of rules.[5] According to this model, the process of cognition involves *mental representation*. Like a computer, the mind is thought to operate by manipulating symbols that represent certain features of the world.[6] This computer model of mental activity was so persuasive and powerful that it dominated all research in cognitive science for over thirty years.

Since the 1940s almost all of neurobiology has been shaped by this idea that the brain is an information-processing device. For example, when studies of the visual cortex showed that certain neurons respond to certain features of perceived objects—velocity, color, contrast, and so on—these feature-specific neurons were seen as picking up visual information from the retina, to be passed on to other areas of the brain for further processing. However, subsequent animal studies made it clear that the association of neurons with specific features can be made only with anesthetized animals in highly controlled internal and external environments. When an animal is studied while it is awake and behaving in more normal surroundings, its neural responses become sensitive to the entire context of the visual stimulus and can no longer be interpreted in terms of stage-by-stage information processing.[7]

The computer model of cognition was finally subjected to serious questioning in the 1970s when the concept of self-organization emerged. The motivation for taking a second look at the dominant hypothesis came from two widely acknowledged deficiencies of the computational view. The first is that information processing is based on sequential rules, applied one at a time; the second is that it is localized, so that an injury to any part of the system results in a serious malfunction of the whole. Both characteristics are in striking contradiction to biological observation. The most ordinary visual tasks, even by tiny insects, are done faster than is physically possible when simulated sequentially; and the resilience of the brain to being damaged without compromising all of its functioning is well-known.

These observations suggested a shift of focus—from symbols to connectivity, from local rules to global coherence, from information processing to the emergent properties of neural networks. With the concurrent development of nonlinear mathematics and models of self-organizing systems, such a shift of focus promised to open up new and intellectually exciting avenues of research. Indeed, by the early 1980s "connectionist" models of neural networks had become very popular.[8] These are models of densely interconnected elements designed to simultaneously carry out millions of operations that generate interesting global, or emergent, properties. As Francisco Varela explains, "The brain is . . . a highly cooperative system: the dense interactions among its components entail that eventually everything going on will be a function of what all the components are doing. . . . As a result the entire system acquires an internal coherence in intricate patterns, even if we cannot say exactly how this occurs."[9]

The Santiago Theory

The Santiago theory of cognition originated in the study of neural networks and, from the very beginning, has been linked to

Maturana's concept of autopoiesis.[10] Cognition, according to Maturana, is the activity involved in the self-generation and self-perpetuation of autopoietic networks. In other words, cognition is the very process of life. "Living systems are cognitive systems," writes Maturana, "and living as a process is a process of cognition."[11] In terms of our three key criteria of living systems—structure, pattern, and process—we can say that the life process consists of all activities involved in the continual embodiment of the system's (autopoietic) pattern of organization in a physical (dissipative) structure.

Since cognition traditionally is defined as the process of knowing, we must be able to describe it in terms of an organism's interactions with its environment. Indeed, this is what the Santiago theory does. The specific phenomenon underlying the process of cognition is structural coupling. As we have seen, an autopoietic system undergoes continual structural changes while preserving its weblike pattern of organization. It couples to its environment *structurally* in other words, through recurrent interactions, each of which triggers structural changes in the system.[12] The living system is autonomous, however. The environment only triggers the structural changes; it does not specify or direct them.

Now, the living system not only specifies these structural changes, it also specifies *which perturbations from the environment trigger them.* This is the key to the Santiago theory of cognition. The structural changes in the system constitute acts of cognition. By specifying which perturbations from the environment trigger its changes, the system "brings forth a world," as Maturana and Varela put it. Cognition, then, is not a representation of an independently existing world, but rather a continual *bringing forth of a world* through the process of living. The interactions of a living system with its environment are cognitive interactions, and the process of living itself is a process of cognition. In the words of Maturana and Varela, "To live is to know."[13]

It is obvious that we are dealing here with a radical expansion of the concept of cognition and, implicitly, the concept of mind. In this new view cognition involves the entire process of life—including perception, emotion, and behavior—and does not necessarily require a brain and a nervous system. Even bacteria perceive certain characteristics of their environment. They sense chemical differences in their surroundings and, accordingly, swim toward sugar and away from acid; they sense and avoid heat, move away from light or toward it, and some bacteria can even detect magnetic fields.[14] Thus even a bacterium brings forth a world—a world of warmth and coldness, of magnetic fields and chemical gradients. In all these cognitive processes perception and action are inseparable, and since the structural changes and associated actions that are triggered in an organism depend on the organism's structure, Francisco Varela describes cognition as "embodied action."[15]

In fact, cognition involves two kinds of activities that are inextricably linked: the maintenance and continuation of autopoiesis and the bringing forth of a world. A living system is a multiply interconnected network whose components are constantly changing, being transformed and replaced by other components. There is great fluidity and flexibility in this network, which allows the system to respond to disturbances, or "stimuli," from the environment in a very special way. Certain disturbances trigger specific structural changes—in other words, changes in the connectivity throughout the network. This is a distributive phenomenon. The entire network responds to a selected disturbance by rearranging its patterns of connectivity.

Different organisms change differently, and over time each organism forms its unique, individual pathway of structural changes in the process of development. Since these structural changes are acts of cognition, development is always associated with learning. In fact, development and learning are two sides of the same coin. Both are expressions of structural coupling.

Not all physical changes in an organism are acts of cognition. When part of a dandelion is eaten by a rabbit, or when an animal is injured in an accident, those structural changes are not specified and directed by the organism; they are not changes of choice and thus not acts of cognition. However, these imposed physical changes are accompanied by other structural changes (perception, response of the immune system, and so forth) that are acts of cognition.

On the other hand, not all disturbances from the environment cause structural changes. Living organisms respond to only a small fraction of the stimuli impinging on them. We all know that we can see or hear phenomena only within a certain range of frequencies; we often do not notice things and events in our environment that do not concern us, and we also know that what we perceive is conditioned largely by our conceptual framework and our cultural context.

In other words, there are many disturbances that do not cause structural changes because they are "foreign" to the system. In this way each living system builds up its own distinctive world according to its own distinctive structure. As Varela puts it, "Mind and world arise together."[16] However, through mutual structural coupling, individual living systems are part of each other's worlds. They communicate with one another and coordinate their behavior.[17] There is an ecology of worlds brought forth by mutually coherent acts of cognition.

In the Santiago theory cognition is an integral part of the way a living organism interacts with its environment. It does not *react* to environmental stimuli through a linear chain of cause and effect, but *responds* with structural changes in its nonlinear, organizationally closed, autopoietic network. This type of response enables the organism to continue its autopoietic organization and thus to continue living in its environment. In other words, the organism's cognitive interaction with its environment is intelligent interaction. From the perspective of the Santiago theory, intelligence is manifest in the richness and flexibility of an organism's structural coupling.

The range of interactions a living system can have with its environment defines its "cognitive domain." Emotions are an integral part of this domain. For example, when we respond to an insult by getting angry, that entire pattern of physiological processes—a red face, faster breathing, trembling, and so on—is part of cognition. In fact, recent research strongly indicates that there is an emotional coloring to every cognitive act.[18]

As the complexity of a living organism increases, so does its cognitive domain. The brain and nervous system, in particular, represent a significant expansion of an organism's cognitive domain, as they greatly increase the range and differentiation of its structural couplings. At a certain level of complexity a living organism couples structurally not only to its environment but also to itself, and thus brings forth not only an external but also an inner world. In human beings the bringing forth of such an inner world is linked intimately to language, thought, and consciousness.[19]

No Representation, No Information

Being part of a unifying conception of life, mind, and consciousness, the Santiago theory of cognition has profound implications for biology, psychology, and philosophy. Among these, its contribution to epistemology, the branch of philosophy that is concerned with the nature of our knowledge about the world, is perhaps its most radical and controversial aspect.

The unique characteristic of the epistemology implied by the Santiago theory is that it takes issue with an idea that is common to most epistemologies but is rarely mentioned explicitly—the idea that cognition is a *representation* of an independently existing world. The computer model of cognition as information processing was merely a specific formulation, based on an erroneous analogy, of the more general idea that the world is pregiven and independent of the observer and that cognition involves mental representations of its objective features inside the cognitive system. The central image, according to Varela, is that of "a cognitive agent parachuted into a pregiven world" and extracting its essential features through a process of representation.[20]

According to the Santiago theory, cognition is not a representation of an independent, pregiven world, but rather a bringing forth of a world. What is brought forth by a particular organism in the process of living is not *the* world but *a* world, one that is always dependent upon the organism's structure. Since individual organisms within a species have more or less the same structure, they bring forth similar worlds. We humans, moreover, share an abstract world of language and thought through which we bring forth our world together.[21]

Maturana and Varela do not maintain that there is a void out there, out of which we create matter. There is a material world, but it does not have any predetermined features. The authors of the Santiago theory do not assert that "nothing exists"; they assert that "no things exist" independent of the process of cognition. There are no objectively existing structures; there is no pregiven territory of which we can make a map—the map making itself brings forth the features of the territory.

We know that cats or birds will see trees, for example, very differently from the way we do, because they perceive light in different frequency ranges. Thus the shapes and textures of the "trees" they bring forth will be different from ours. When we see a tree, we are not inventing reality. But the ways in which we delineate objects and identify patterns out of the multitude of sensory inputs we receive depends on our physical constitution. As Maturana and Varela would say, the ways in which we can couple structurally to our environment, and thus the world we bring forth, depend on our own structure.

Together with the idea of mental representations of an independent world, the Santiago theory also rejects the idea of information as some objective features of that independently existing world. In Varela's words:

> We must call into question the idea that the world is pregiven and that cognition is representation. In cognitive science, this means that we must call into question the idea that information exists ready-made in the world and that it is extracted by a cognitive system.[22]

The rejection of representation and of information as being relevant to the process of knowing are both difficult to accept, because we use both concepts constantly. The symbols of our language, both spoken and written, are representations of things and ideas; and in our daily lives we consider facts such as the time of day, the date, the weather report, or the telephone number of a friend as pieces of information that are relevant to us. In fact, our whole era has often been called the "information age." So how can

Maturana and Varela claim that there is no information in the process of cognition?

To understand that seemingly puzzling assertion, we must remember that for human beings cognition involves language, abstract thinking, and symbolic concepts that are not available to other species. The ability to abstract is a key characteristic of human consciousness, as we shall see, and because of that ability we can and do use mental representations, symbols, and information. However, these are not characteristics of the general process of cognition that is common to all living systems. Although human beings frequently use mental representations and information, our cognitive process is not based on them.

To gain a proper perspective on these ideas, it is very instructive to take a closer look at what is meant by "information." The conventional view is that information is somehow "lying out there" to be picked up by the brain. However, such a piece of information is a quantity, name, or short statement that we have abstracted from a whole network of relationships, a context, in which it is embedded and which gives it meaning. Whenever such a "fact" is embedded in a stable context that we encounter with great regularity, we can abstract it from that context, associate it with the meaning inherent in the context, and call it "information." We are so used to these abstractions that we tend to believe that meaning resides in the piece of information rather than in the context from which it has been abstracted.

For example, there is nothing "informative" in the color red, except that, when embedded in a cultural network of conventions and in the technological network of city traffic, it is associated with stopping at an intersection. If people from a very different culture came to one of our cities and saw a red traffic light, it might not mean anything to them. There would be no information conveyed. Similarly, the time of day and the date are abstracted from a complex context of concepts and ideas, including a model of the solar system, astronomical observations, and cultural conventions.

The same considerations apply to the genetic information encoded in DNA. As Varela explains, the notion of a genetic code has been abstracted from an underlying metabolic network in which the meaning of the code is embedded:

> For many years biologists considered protein sequences as being instructions coded in the DNA. It is clear, however, that DNA triplets are capable of predictably specifying an amino acid in a protein if and only if they are embedded in the cell's metabolism, that is, in the thousands of enzymatic regulations in a complex chemical network. It is only because of the emergent regularities of such a network as a whole that we can bracket out this metabolic background and thus treat triplets as codes for amino acids.[23]

Maturana and Bateson

Maturana's rejection of the idea that cognition involves a mental representation of an independent world is the key difference between his conception of the process of knowing and that of Gregory Bateson. Maturana and Bateson, around the same time, independently hit upon the revolutionary idea of identifying the process of knowing with the process of life.[24] But they approached it in very different ways—Bateson from a deep intuition of the nature of mind and life, honed by careful observations of the living world; Maturana from his attempts to define a pattern of organization that is characteristic of all living systems, based on his research in neuroscience.

Bateson, working alone, refined his "criteria of mental process" over the years but never developed them into a theory of living systems. Maturana, by contrast, collaborated with other scientists to develop a theory of "the organization of the living," which provides the theoretical framework for understanding the process of cognition as the process of life. As social scientist Paul Dell put it in his extensive paper, "Understanding Bateson and Maturana," Bateson concentrated exclusively on epistemology (the nature of knowledge) at the expense of dealing with ontology (the nature of existence):

> Ontology constitutes "the road not taken" in Bateson's thinking. . . . Bateson's epistemology has no ontology upon which to found

itself. . . . It is my contention that Maturana's work contains the ontology that Bateson never developed.[25]

An examination of Bateson's criteria of mental process shows that they cover both the structure aspect and the pattern aspect of living systems, which may be the reason why many of Bateson's students found them rather confusing. A close reading of the criteria also reveals the underlying belief that cognition involves mental representations of the world's objective features inside the cognitive system.[26]

Bateson and Maturana independently created a revolutionary concept of mind that is rooted in cybernetics, a tradition that Bateson helped to develop in the 1940s. Perhaps it was because of his intimate involvement with cybernetic ideas during the time of their genesis that Bateson never transcended the computer model of cognition. Maturana, by contrast, left that model behind and developed a theory that views cognition as the act of "bringing forth a world" and consciousness as being closely associated with language and abstraction.

Computers Revisited

In the previous pages I have repeatedly emphasized the differences between the Santiago theory and the computational model of cognition developed in cybernetics. It might now be useful to take another look at computers in light of our new understanding of cognition, in order to dispel some of the confusion surrounding "computer intelligence."

A computer processes information, which means that it manipulates symbols based on certain rules. The symbols are distinct elements fed into the computer from outside, and during the information processing there is no change in the structure of the machine. The physical structure of the computer is fixed, determined by its design and construction.

The nervous system of a living organism works very differently. As we have seen, it interacts with its environment by continually modulating its structure, so that at any moment its physical structure is a record of previous structural changes. The nervous system does not process information from the outside world but, on the contrary, *brings forth* a world in the process of cognition.

Human cognition involves language and abstract thinking, and thus symbols and mental representations, but abstract thought is only a small part of human cognition and generally is not the basis for our everyday decisions and actions. Human decisions are never completely rational but are always colored by emotions, and human thought is always embedded in the bodily sensations and processes that contribute to the full spectrum of cognition.

As computer scientists Terry Winograd and Fernando Flores point out in their book, *Understanding Computers and Cognition,* rational thought filters out most of that cognitive spectrum and, in so doing, creates a "blindness of abstraction." Like blinders, the terms we adopt to express ourselves limit the range of our view. In a computer program, Winograd and Flores explain, various goals and tasks are formulated in terms of a limited collection of objects, properties, and operations, a collection that embodies the blindness that comes with the abstractions involved in creating the program. However:

> There are restricted task domains in which this blindness does not preclude a behavior that appears intelligent. For example, many games are amenable to a direct application of . . . techniques [that can] produce a program that outplays human opponents. . . . These are areas in which the identification of the relevant features is straightforward and the nature of solutions is clearcut.[27]

A lot of confusion is caused by the fact that computer scientists use words such as "intelligence," "memory," and "language" to describe computers, thus implying that these expressions refer to the human phenomena we know well from experience. This is a serious misunderstanding. For example, the very essence of intelligence is to act appropriately when a problem is not clearly defined and solutions are not evident. Intelligent human behavior in such situations is based on common sense, accumulated from lived experience. Common sense, however, is not available to computers because of their blindness of abstraction and the intrinsic limitations of formal operations, and therefore it is impossible to program computers to be intelligent.[28]

Since the early days of artificial intelligence one of the greatest challenges has been to program a computer to understand human language. But after several decades of frustrating work on this problem, researchers in AI are beginning to realize that their efforts are bound to remain futile, that computers cannot understand human language in a significant sense.[29] The reason is that language is embedded in a web of social and cultural conventions that provides an unspoken context of meaning. We understand this context because it is common sense to us, but a computer cannot be programmed with common sense and therefore does not understand language.

This point can be illustrated with many simple examples, such as this text used by Terry Winograd: "Tommy had just been given a new set of blocks. He was opening the box when he saw Jimmy coming in." As Winograd explains, a computer would have no clue as to what is in the box, but we assume immediately that it contains Tommy's new blocks. We do so because we know that gifts often come in boxes and that opening the box is the proper thing to do. Most important, we assume that the two sentences in the text are connected, whereas the computer sees no reason to connect the box with the blocks. In other words, our interpretation of this simple text is based on several commonsense assumptions and expectations that are unavailable to the computer.[30]

The fact that a computer cannot understand language does not mean that it cannot be programmed to recognize and manipulate simple linguistic structures. Indeed, much progress has been made in this area in recent years. Computers can now recognize a few hundred words and phrases, and this basic vocabulary keeps expanding. Thus machines are used increasingly to interact with people through the structures of human language to carry out limited tasks. For example, I can call my bank for information about my checking account, and a computer, if prompted by a sequence of codes, will give me the amount of my balance, the number and amounts of recent checks and deposits, and so on. This interaction, which involves a combination of simple spoken words and punched-in numbers, is very convenient and useful without implying in any way that the bank's computer understands human language.

Unfortunately there is a striking dissonance between serious critical assessments of AI and the optimistic projections of the computer industry, which are strongly motivated by commercial interests. The most recent wave of enthusiastic pronouncements has come from the so-called fifth generation project launched in Japan. An analysis of its grandiose goals suggests, however, that they are as unrealistic as similar earlier projections, even though the program is likely to produce numerous useful spinoffs.[31]

The centerpiece of the fifth generation project and of other similar research projects is the development of so-called expert systems, to be designed to rival the performance of human experts in certain tasks. This is again an unfortunate use of terminology, as Winograd and Flores point out:

> Calling a program an "expert" is misleading in exactly the same way as calling it "intelligent" or saying it "understands." The misrepresentation may be useful for those who are trying to get research funding or sell such programs, but it can lead to inappropriate expectations by those who attempt to use them.[32]

In the mid-1980s philosopher Hubert Dreyfus and computer scientist Stuart Dreyfus undertook a thorough study of human expertise and contrasted it with computer expert systems. They found that

> . . . one has to abandon the traditional view that a beginner starts with specific cases and, as he becomes more proficient, abstracts and interiorizes more and more sophisticated rules. . . . Skill acquisition moves in just the opposite direction—from abstract rules to particular cases. It seems that a beginner makes inferences using rules and facts just like a heuristically programmed computer, but with talent and a great deal of involved experience the beginner develops into an expert who intuitively sees what to do without applying rules.[33]

This observation explains why expert systems never perform as well as experienced human experts, who do not operate by applying a sequence of rules, but act on the basis of their intuitive grasp of an entire constellation of facts. Dreyfus and Dreyfus also noted that in practice, expert systems are designed by asking human experts for the relevant rules. When this is done the experts tend to state the rules they remember from the time when they were beginners, but which they stopped using when they became experts. If these rules are programmed into a computer, the resulting expert system will outperform a human beginner using the same rules but can never rival a true expert.

Cognitive Immunology

Some of the most important practical applications of the Santiago theory are those that are likely to arise from its impact on neuroscience and immunology. As mentioned previously, the new view of cognition greatly clarifies the age-old puzzle about the relationship between mind and brain. Mind is not a thing but a process—the process of cognition, which is identified with the process of life. The brain is a specific structure through which this process operates. Thus the relationship between mind and brain is one between process and structure.

The brain is by no means the only structure involved in the process of cognition. In the human organism, as in the organisms of all vertebrates, the immune system is increasingly being recognized as a network that is as complex and interconnected as the nervous system and serves equally important coordinating functions. Classical immunology sees the immune system as the body's defense system, outwardly directed and often described in terms of military metaphors—armies of white blood cells, generals, soldiers, and so on. Recent discoveries by Francisco Varela and his colleagues at the University of Paris are seriously challenging this conception.[34] In fact, some researchers now believe that the classical view with its military metaphors has been one of the main stumbling blocks in our understanding of autoimmune diseases such as AIDS.

Instead of being concentrated and interconnected through anatomical structures like the nervous system, the immune system is dispersed in the lymph fluid, permeating every single tissue. Its components—a class of cells called lymphocytes, popularly known as white blood cells—move around very rapidly and bind chemically to each other. The lymphocytes are an extremely diverse group of cells. Each type is distinguished by specific molecular markers, called "antibodies," sticking out from their surfaces. The human body contains billions of different types of white blood cells, with an enormous ability to bind chemically to any molecular profile in their environment.

According to traditional immunology, the lymphocytes identify an intruding agent, the antibodies attach themselves to it and, by doing so, neutralize it. This sequence implies that the white blood cells recognize foreign molecular profiles. Closer examination shows that it also implies some form of learning and memory. In classical immunology, however, these terms are used purely metaphorically, without allowing for any actual cognitive processes.

Recent research has shown that under normal conditions the antibodies circulating in the body bind to many (if not all) types of cells, including themselves. The entire system looks much more like a network, more like people talking to each other, than soldiers out looking for an enemy. Gradually immunologists have been forced to shift their perception from an immune *system* to an immune *network*.

This shift in perception presents a big problem for the classical view. If the immune system is a network whose components bind to each other, and if antibodies are meant to eliminate whatever they bind to, we should all be destroying ourselves. Obviously we are not. The immune system seems to be able to distinguish between its own body's cells and foreign agents, between self and nonself. But since, in the classical view, for an antibody to recognize a foreign agent means binding to it chemically and thereby neutralizing it, it remains mysterious how the immune system can recognize its own cells without neutralizing (that is, functionally destroying) them.

Furthermore, from the traditional point of view an immune

system will develop only when there are outside disturbances to which it can respond. If there is no attack, no antibodies will be developed. Recent experiments have shown, however, that even animals that are completely sheltered from disease-causing agents still develop full-blown immune systems. From the new point of view this is natural, because the immune system's main function is not to respond to outside challenges, but to relate to itself.[35]

Varela and his colleagues argue that the immune system needs to be understood as an autonomous, cognitive network, which is responsible for the body's "molecular identity." By interacting with one another and with the other body cells, the lymphocytes continually regulate the number of cells and their molecular profiles. Rather than merely reacting against foreign agents, the immune system serves the important function of regulating the organism's cellular and molecular repertoire. As Francisco Varela and immunologist Antonio Coutinho explain, "The mutual dance between immune system and body . . . allows the body to have a changing and plastic identity throughout its life and its multiple encounters."[36]

From the perspective of the Santiago theory, the cognitive activity of the immune system results from its structural coupling to its environment. When foreign molecules enter the body, they perturb the immune network, triggering structural changes. The resulting response is not automatic destruction of the foreign molecules, but regulation of their levels within the context of the system's other regulatory activities. The response will vary and will depend upon the entire context of the network.

When immunologists inject large amounts of a foreign agent into the body, as they do in standard animal experiments, the immune system reacts with the massive defensive response described in the classical theory. However, as Varela and Coutinho point out, this is a highly contrived laboratory situation. In its natural surroundings an animal does not receive large amounts of harmful substances. The small amounts that do enter its body are incorporated naturally into the ongoing regulatory activities of its immune network.

With this understanding of the immune system as a cognitive, self-organizing, and self-regulating network, the puzzle of the self/non-self distinction is easily resolved. The immune system simply does not and need not distinguish between body cells and foreign agents, because both are subject to the same regulatory processes. However, when the invading foreign agents are so massive that they cannot be incorporated into the regulatory network, as for example in the case of infections, they will trigger specific mechanisms in the immune system that mount a defensive response.

Research has shown that this well-known immune response involves quasi-automatic mechanisms that are largely independent of the network's cognitive activities.[37] Traditionally immunology has been concerned almost exclusively with such "reflexive" immune activity. To limit ourselves to these studies would correspond to limiting brain research to the study of reflexes. Defensive immune activity is very important, but in the new view it is a secondary effect of the much more central cognitive activity of the immune system, which maintains the body's molecular identity.

The field of cognitive immunology is still in its infancy, and the self-organizing properties of immune networks are by no means well understood. However, some of the scientists active in this growing field of research have already begun to speculate about exciting clinical applications to the treatment of autoimmune diseases.[38] Future therapeutic strategies are likely to be based on the understanding that autoimmune diseases reflect a failure in the cognitive operation of the immune network and may involve various novel techniques designed to reinforce the network by boosting its connectivity.

Such techniques, however, will require a much deeper understanding of the rich dynamics of immune networks before they can be applied effectively. In the long run the discoveries of cognitive immunology promise to be tremendously important for the whole field of health and healing. In Varela's opinion a sophisticated psychosomatic ("mind-body") view of health will not develop until we understand the nervous system and the immune system as two interacting cognitive systems, two "brains" in continuous conversation.[39]

A Psychosomatic Network

A crucial link in this picture was provided in the mid-1980s by neuroscientist Candace Pert and her colleagues at the National Institute of Mental Health in Maryland. These researchers identified a group of molecules, called peptides, as the molecular messengers that facilitate the conversation between the nervous system and the immune system. In fact, Pert and her colleagues have found that these messengers interconnect three distinct systems—the nervous system, the immune system, and the endocrine system—into one single network.

In the traditional view these three systems are separate and serve different functions. The *nervous system,* consisting of the brain and of a network of nerve cells throughout the body, is the seat of memory, thought, and emotion. The *endocrine system,* consisting of the glands and the hormones, is the body's main regulatory system, controlling and integrating various bodily functions. The *immune system,* consisting of the spleen, the bone marrow, the lymph nodes, and the immune cells circulating through the body, is the body's defense system, responsible for tissue integrity and controlling wound healing and tissue-repair mechanisms.

In accord with this separation the three systems are studied in three separate disciplines—neuroscience, endocrinology, and immunology. However, the recent peptide research has shown in dramatic ways that these conceptual separations are merely historical artifacts that can no longer be maintained. According to Candace Pert, the three systems must be seen as forming a single psychosomatic network.[40]

The peptides, a family of sixty to seventy macromolecules, were originally studied in other contexts and were given different names—hormones, neurotransmitters, endorphins, growth factors, and so on. It took many years to recognize that they are a single family of molecular messengers. These messengers are short chains of amino acids that attach themselves to specific receptors, which exist in abundance on the surfaces of all cells of the body. By interlinking immune cells, glands, and brain cells, peptides form a psychosomatic network extending throughout the entire organism. Peptides are the biochemical manifestation of emotions; they play a crucial role in the coordinating activities of the immune system; they interlink and integrate mental, emotional, and biological activities.

A dramatic change of perception began in the early eighties with the controversial discovery that certain hormones, which were supposed to be produced by glands, are peptides and are also produced and stored in the brain. Conversely scientists found that a type of neurotransmitters called endorphins, which were thought to be produced only in the brain, are also produced in immune cells. As more and more peptide receptors were identified, it turned out that virtually any known peptide is produced in the brain *and* in various parts of the body. Thus Candace Pert declares: "I can no longer make a strong distinction between the brain and the body."[41]

In the nervous system peptides are produced in nerve cells and then travel down the axons (the long branches of nerve cells) to be stored in little balls at the bottom, where they wait for the right signals to release them. These peptides play a vital role in communications throughout the nervous system. Traditionally it was thought that the transfer of all nervous impulses occurs across the gaps, called "synapses," between adjacent nerve cells. But this mechanism turns out to be of limited importance, being used mainly for muscle contraction. Most of the signals that come from the brain are transmitted via peptides emitted by nerve cells. By attaching themselves to receptors far away from the nerve cells in which they originated, these peptides act not only throughout the entire nervous system, but also in other parts of the body.

In the immune system the white blood cells not only have receptors for all the peptides, they also *make* peptides themselves. Peptides control the migration patterns of immune cells and all their vital functions. This discovery, like those in cognitive immunology, is likely to generate exciting therapeutic applications. Indeed, Pert and her team recently discovered a new treatment for AIDS, called Peptide T, that holds great promise.[42] The scientists hypothesize that AIDS is rooted in a disruption of peptide communication. They discovered that the HIV enters cells through particular peptide receptors, thereby interfering with the functions of the entire network, and they designed a protective peptide that attaches itself to these receptors and thus blocks the action of HIV. (Peptides occur naturally in the body but can be designed and synthesized as well.) Peptide T mimics the action of a naturally occurring peptide and is therefore completely nontoxic, in contrast with all other AIDS medications. The drug is currently going through a series of clinical trials. If it proves to be effective, it could have a revolutionary impact on the treatment of AIDS.

Another fascinating aspect of the newly recognized psychosomatic network is the discovery that peptides are the biochemical manifestation of emotions. Most peptides, if not all, alter behavior and mood states, and scientists now hypothesize that each peptide may evoke a unique emotional "tone." The entire group of sixty to seventy peptides may constitute a universal biochemical language of emotions.

Traditionally neuroscientists have associated emotions with specific areas in the brain, notably the limbic system. This is indeed correct. The limbic system turns out to be highly enriched with peptides. However, it is not the only part of the body where peptide receptors are concentrated. For example, the entire intestine is lined with peptide receptors. This is why we have "gut feelings." We literally feel our emotions in our gut.

If it is true that each peptide mediates a particular emotional state, this would mean that all sensory perceptions, all thoughts, and, in fact, all bodily functions are emotionally colored, because they all involve peptides. Indeed, scientists have observed that the nodal points of the central nervous system, which connect the sensory organs with the brain, are enriched with peptide receptors that filter and prioritize sensory perceptions. In other words, all our perceptions and thoughts are colored by emotions. This, of course, is also our common experience.

The discovery of this psychosomatic network implies that the nervous system is not hierarchically structured, as had been believed before. As Candace Pert puts it, "White blood cells are bits of the brain floating around in the body."[43] Ultimately this implies that cognition is a phenomenon that expands throughout the organism, operating through an intricate chemical network of peptides that integrates our mental, emotional, and biological activities.

12

Knowing That We Know

Identifying cognition with the full process of life—including perceptions, emotions, and behavior—and understanding it as a process that involves neither a transfer of information nor mental representations of an outside world requires a radical expansion of our scientific and philosophical frameworks. One of the reasons why this view of mind and cognition is so difficult to accept is that it runs counter to our everyday intuition and experience. As human beings we frequently use the concept of information and we constantly make mental representations of the people and objects in our environment.

However, these are specific characteristics of human cognition that result from our ability to abstract, which is a key characteristic of human consciousness. For a thorough understanding of the general process of cognition in living systems it is thus important to understand how human consciousness, with its abstract thought and symbolic concepts, arises out of the cognitive process that is common to all living organisms.

In the following pages I shall use the term "consciousness" to describe the level of mind, or cognition, that is characterized by self-awareness. Awareness of the environment, according to the Santiago theory, is a property of cognition at all levels of life. Self-awareness, as far as we know, is manifest only in higher animals and fully unfolds in the human mind. As humans we are not only aware of our environment, we are also aware of ourselves and our inner world. In other words, we are aware that we are aware. We not only know; we also know that we know. It is this special faculty of self-awareness that I refer to when I use the term "consciousness."

Language and Communication

In the Santiago theory self-awareness is viewed as being tied closely to language, and the understanding of language is approached through a careful analysis of communication. This approach to understanding consciousness has been pioneered by Humberto Maturana.[1]

Communication, according to Maturana, is not a transmission of information, but rather a *coordination of behavior* among living organisms through mutual structural coupling. Such mutual coordination of behavior is the key characteristic of communication for all living organisms, with or without nervous systems, and it becomes more and more subtle and elaborate with nervous systems of increasing complexity.

Birdsongs are among the most beautiful kinds of nonhuman communication, which Maturana illustrates with the stunning example of a particular mating song used by African parrots. These birds often live in dense forests with hardly any possibility of visual contact. In this environment parrot couples form and coordinate their mating ritual by producing a common song. To the casual listener it seems that each bird is singing a full melody, but closer inspection shows that this melody is actually a duet in which the two birds alternatively expand upon each other's phrases.

The whole melody is unique to each couple and is not passed on to their offspring. In each generation new couples will produce their own characteristic melodies in their mating rituals. In Maturana's words:

> In this case (unlike with many other birds), the vocal coordination of behavior in the singing couple is an ontogenic [i.e. developmental] phenomenon. . . . The particular melody of each couple in this species of bird is unique to its history of coupling.[2]

This is a clear and beautiful example of Maturana's observation that communication is essentially a coordination of behavior. In other cases we may be more tempted to describe communication in semantic terms—that is, in terms of an exchange of information that carries some meaning. However, according to Maturana, such semantic descriptions are projections by the human observer. In reality the coordination of behavior is determined not by meaning but by the dynamics of structural coupling.

Animal behavior may be inborn ("instinctive") or learned, and accordingly we can distinguish between instinctive and learned communication. Maturana calls the learned communicative behavior "linguistic." Although it is not yet language, it shares with language the characteristic feature that the same coordination of behavior may be achieved by different types of interactions. Like different languages in human communication, different kinds of structural couplings, learned along different developmental paths, may result in the same coordination of behavior. Indeed, in Maturana's view such linguistic behavior is the basis for language.

Linguistic communication requires a nervous system of considerable complexity, because it involves quite a lot of complex learning. For example, when honeybees indicate the location of specific flowers to each other by dancing out intricate patterns, those dances are partly based on instinctive behavior and partly learned. The linguistic (or learned) aspects of the dance are specific to the context and social history of the beehive. Bees from different hives dance in different "dialects," so to speak.

Even very intricate forms of linguistic communication, such as the so-called language of bees, are not yet language. According to Maturana, language arises when there is *communication about communication.* In other words, the process of "languaging," as Maturana calls it, takes place when there is a coordination of coordinations of behavior. Maturana likes to illustrate this meaning of language with a hypothetical communication between a cat and her owner.[3]

Suppose that every morning my cat meows and runs to the refrigerator. I follow her, take out some milk, and pour it into a bowl, and the cat begins to lap it up. That is communication—a coordination of behavior through recurrent mutual interactions, or mutual structural coupling. Now suppose that one morning I don't follow the meowing cat because I know that I've run out of milk. If the cat were somehow able to communicate to me something like "Hey, I've now meowed three times; where is my milk?" that would be language. Her reference to her previous meowing would constitute a communication about a communication, and thus, according to Maturana's definition, would qualify as language.

Cats are unable to use language in that sense, but higher apes may well be able to do so. In a series of well-publicized experiments American psychologists showed that chimpanzees are able not only to learn many standard signs of a sign language, but to create new expressions by combining various signs.[4] Thus one of the chimps, named Lucy, invented several sign combinations: "fruit-drink" for watermelon, "food-cry-strong" for radish, and "open-drink-eat" for refrigerator.

One day, when Lucy got very upset upon seeing that her human "parents" were getting ready to leave, she turned to them and signed "Lucy cry." By making this statement about her crying, she evidently communicated something about a communication. "It seems to us," write Maturana and Varela, "that, at this point, Lucy is languaging."[5]

Although some primates seem to have the potential of communicating in sign language, their linguistic domain is extremely limited and does not come anywhere near the richness of human language. In human language a vast space is opened up in which words serve as tokens for the linguistic coordination of actions and are also used to create the notion of objects. For example, at a picnic we can use words as linguistic distinctions to coordinate our actions of putting a tablecloth and food on a tree stump. In addition, we can also refer to those linguistic distinctions (in other words, make a distinction of distinctions) by using the word "table" and thus bringing forth an object.

Objects, then, in Maturana's view, are linguistic distinctions of linguistic distinctions, and once we have objects we can create abstract concepts—the height of our table, for example—by making distinctions of distinctions of distinctions, and so forth. Using Bateson's terminology, we could say that a hierarchy of logical types emerges with human language.[6]

Languaging

Our linguistic distinctions, moreover, are not isolated but exist "in the network of structural couplings that we continually weave through [languaging]."[7] Meaning arises as a pattern of relationships among these linguistic distinctions, and thus we exist in a "semantic domain" created by our languaging. Finally, self-awareness arises when we use the notion of an object and the associated abstract concepts to describe ourselves. Thus the linguistic domain of human beings expands further to include reflection and consciousness.

The uniqueness of being human lies in our ability to continually weave the linguistic network in which we are embedded. To be human is to exist in language. In language we coordinate our behavior, and together in language we bring forth our world. "The world everyone sees," write Maturana and Varela, "is not *the* world but *a* world, which we bring forth with others."[8] This human world centrally includes our inner world of abstract thought, concepts, symbols, mental representations, and self-awareness. To be human is to be endowed with reflective consciousness: "As we know how we know, we bring forth ourselves."[9]

In a human conversation our inner world of concepts and ideas, our emotions, and our body movements become tightly linked in a complex choreography of behavioral coordination. Film analyses have shown that every conversation involves a subtle and largely unconscious dance in which the detailed sequence of speech patterns is precisely synchronized not only with minute movements of the speaker's body, but also with corresponding movements of the listener. Both partners are locked into this precisely synchronized sequence of rhythmic movements, and the linguistic coordination of their mutually triggered gestures lasts as long as they remain involved in their conversation.[10]

Maturana's theory of consciousness differs fundamentally from most others because of its emphasis on language and communication. From the perspective of the Santiago theory, the currently fashionable attempts to explain human consciousness in terms of quantum effects in the brain or other neurophysiological processes are all bound to fail. Self-awareness and the unfolding of our inner world of concepts and ideas are not only inaccessible to explanations in terms of physics and chemistry; they cannot even be understood through the biology or psychology of a single organism. According to Maturana, we can understand human consciousness only through language and the whole social context in which it is embedded. As its Latin root—*con-scire* ("knowing together")—might indicate, consciousness is essentially a social phenomenon.

It is also instructive to compare the notion of bringing forth a world with the ancient Indian concept of *maya.* The original meaning of *maya* in early Hindu mythology is the "magic creative power" by which the world is created in the divine play of Brahman.[11] The myriad forms we perceive are all brought forth by the divine actor and magician, and the dynamic force of the play is *karma,* which literally means "action."

Over the centuries the word *maya*—one of the most important terms in Indian philosophy—changed its meaning. From the creative power of Brahman it came to signify the psychological state of anybody under the spell of the magic play. As long as we confuse the material forms of the play with objective reality, without perceiving the unity of Brahman underlying all these forms, we are under the spell of *maya.*

Hinduism denies the existence of an objective reality. As in the Santiago theory, the objects we perceive are brought forth through action. However, the process of bringing forth the world occurs on a cosmic scale rather than at the human level of cognition. The world brought forth in Hindu mythology is not *a* world for a particular human society bound together by language and culture, but *the* world of the magic divine play that holds us all under its spell.

Primary States of Consciousness

In recent years Francisco Varela has been following another approach to consciousness that, he hopes, may add an additional dimension to Maturana's theory. His basic hypothesis is that there is a form of primary consciousness in all higher vertebrates that is not yet self-reflective but involves the experience of a "unitary mental space," or "mental state."

Numerous recent experiments with animals and humans have shown that this mental space is composed of many dimensions—in other words, it is created by many different brain functions—and yet it is a single coherent experience. For example, when the smell of a perfume evokes a pleasant or unpleasant sensation, one experiences a single, coherent mental state composed of sensory perceptions, memories, and emotions. The experience is not constant, as we well know, and may be extremely short. Mental states are transitory, continually arising and subsiding. However, it does not seem possible to experience them without some finite span of duration. Another important observation is that the experiential state is always "embodied"—that is, embedded in a particular field of sensation. In fact, most mental states seem to have a dominant sensation that colors the entire experience.

Varela recently published a paper in which he sets forth his basic hypothesis and proposes a specific neural mechanism for the constitution of primary states of consciousness in all higher vertebrates.[12] The key idea is that transitory experiential states are created by a resonance phenomenon known as "phase locking," in which different brain regions are interconnected in such a way that all their neurons fire in synchrony. Through this synchronization of neural activity, temporary "cell assemblies" are formed, which may consist of widely dispersed neural circuits.

According to Varela's hypothesis, each cognitive experience is based on a specific cell assembly, in which many different neural activities—associated with sensory perception, emotions, memory, bodily movements, and so on—are unified into a transient but coherent ensemble of oscillating neurons. The fact that neural circuits tend to oscillate rhythmically is well-known to neuroscientists, and recent research has shown that these oscillations are not restricted to the cerebral cortex but occur at various levels in the nervous system.

The numerous experiments cited by Varela in support of his hypothesis indicate that cognitive experiential states are created by the synchronization of fast oscillations in the gamma and beta range that tend to arise and subside quickly. Each phase locking is associated with a characteristic relaxation time, which accounts for the minimum duration of the experience.

Varela's hypothesis establishes a neurological basis for the distinction between conscious and unconscious cognition, which neuroscientists have been looking for ever since Sigmund Freud discovered the human unconscious.[13] According to Varela, the primary conscious experience, common to all higher vertebrates, is not located in a specific part of the brain, nor can it be identified in terms of specific neural structures. It is the manifestation of a particular cognitive process—a transient synchronization of diverse, rhythmically oscillating neural circuits.

The Human Condition

Human beings evolved from the upright walking "Southern apes" (genus *Australopithecus*) around two million years ago. The transition from apes to humans, as we have learned in an earlier chapter, was driven by two distinct developments: the helplessness of prematurely born infants, which required supportive families and communities, and the freedom of the hands to make and use tools, which stimulated brain growth and may have contributed to the evolution of language.[14]

Maturana's theory of language and consciousness allows us to interlink these two evolutionary drives. Since language results in a very sophisticated and effective coordination of behavior, the evolution of language allowed the early human beings to greatly increase their cooperative activities and to develop families, communities, and tribes that gave them tremendous evolutionary advantages. The crucial role of language in human evolution was not the ability to exchange ideas, but the increased ability to cooperate.

As the diversity and richness of our human relationships increased, our humanity—our language, art, thought, and culture—unfolded accordingly. At the same time, we also developed the ability of abstract thinking, of bringing forth an inner world of concepts, objects, and images of ourselves. Gradually, as this inner world became ever more diverse and complex, we began to lose touch with nature and became ever more fragmented personalities.

Thus arose the tension between wholeness and fragmentation, between body and soul, which has been identified as the essence of the human condition by poets, philosophers, and mystics throughout the ages. Human consciousness has brought forth not only the Chauvet cave paintings, the Bhagavad Gita, the Brandenburg Concertos, and the theory of relativity, but also slavery, witch burnings, the Holocaust, and the bombing of Hiroshima. Among all the species, we are the only ones that kill their own kind in pursuit of religion, free markets, patriotism, and other abstract ideas.

Buddhist philosophy contains some of the most lucid expositions of the human condition and its roots in language and consciousness.[15] Existential human suffering arises, in the Buddhist view, when we cling to fixed forms and categories created by the mind instead of accepting the impermanent and transitory nature of all things. The Buddha taught that all fixed forms—things, events, people, or ideas—are nothing but *maya*. Like the Vedic seers and sages, he used this ancient Indian concept but brought it down from the cosmic level it occupies in Hinduism, connecting it with the process of human cognition and thus giving it a fresh, almost psychotherapeutic interpretation.[16] Out of ignorance *(avidya),* we divide the perceived world into separate objects that we see as firm and permanent, but which are really transient and ever-changing. Trying to cling to our rigid categories instead of realizing the fluidity of life, we are bound to experience frustration after frustration.

The Buddhist doctrine of impermanence includes the notion that there is no self—no persistent subject of our varying experiences. It holds that the idea of a separate, individual self is an illusion, just another form of *maya,* an intellectual concept that has no reality. To cling to this idea of a separate self leads to the same pain and suffering *(duhkha)* as the adherence to any other fixed category of thought.

Cognitive science has arrived at exactly the same position.[17] According to the Santiago theory, we bring forth the self just as we bring forth objects. Our self, or ego, does not have any independent existence but is a result of our internal structural coupling. A detailed analysis of the belief in an independent, fixed self and the resulting "Cartesian anxiety" leads Francisco Varela and his colleagues to the following conclusion:

> Our grasping after an inner ground is the essence of ego-self and is the source of continuous frustration. . . . This grasping after an inner ground is itself a moment in a larger pattern of grasping that includes our clinging to an outer ground in the form of the idea of a pregiven and independent world. In other words, our grasping after a ground, whether inner or outer, is the deep source of frustration and anxiety.[18]

This, then, is the crux of the human condition. We are autonomous individuals, shaped by our own history of structural changes. We are self-aware, aware of our individual identity—and yet when we look for an independent self within our world of experience we cannot find any such entity.

The origin of our dilemma lies in our tendency to create the abstractions of separate objects, including a separate self, and then to believe that they belong to an objective, independently existing reality. To overcome our Cartesian anxiety, we need to think systemically, shifting our conceptual focus from objects to relationships. Only then can we realize that identity, individuality, and autonomy do not imply separateness and indcpendence. As Lynn

Margulis and Dorion Sagan remind us, "Independence is a political, not a scientific, term."[19]

The power of abstract thinking has led us to treat the natural environment—the web of life—as if it consisted of separate parts, to be exploited by different interest groups. Moreover, we have extended this fragmented view to our human society, dividing it into different nations, races, religious and political groups. The belief that all these fragments—in ourselves, in our environment, and in our society—are really separate has alienated us from nature and from our fellow human beings and thus has diminished us. To regain our full humanity, we have to regain our experience of connectedness with the entire web of life. This reconnecting, *religio* in Latin, is the very essence of the spiritual grounding of deep ecology.

Epilogue: Ecological Literacy

Reconnecting with the web of life means building and nurturing sustainable communities in which we can satisfy our needs and aspirations without diminishing the chances of future generations. For this task we can learn valuable lessons from the study of ecosystems, which *are* sustainable communities of plants, animals, and microorganisms. To understand these lessons, we need to learn the basic principles of ecology. We need to become, as it were, ecologically literate.[1] Being ecologically literate, or "ecoliterate," means understanding the principles of organization of ecological communities (ecosystems) and using those principles for creating sustainable human communities. We need to revitalize our communities—including our educational communities, business communities, and political communities—so that the principles of ecology become manifest in them as principles of education, management, and politics.[2]

The theory of living systems discussed in this book provides a conceptual framework for the link between ecological communities and human communities. Both are living systems that exhibit the same basic principles of organization. They are networks that are organizationally closed, but open to the flows of energy and resources; their structures are determined by their histories of structural changes; they are intelligent because of the cognitive dimensions inherent in the processes of life.

Of course, there are many differences between ecosystems and human communities. There is no self-awareness in ecosystems, no language, no consciousness, and no culture; and therefore no justice or democracy; but also no greed or dishonesty. We cannot learn anything about those human values and shortcomings from ecosystems. But what we *can* learn and must learn from them is how to live sustainably. During more than three billion years of evolution the planet's ecosystems have organized themselves in subtle and complex ways so as to maximize sustainability. This wisdom of nature is the essence of ecoliteracy.

Based on the understanding of ecosystems as autopoietic networks and dissipative structures, we can formulate a set of principles of organization that may be identified as the basic principles of ecology and use them as guidelines to build sustainable human communities.

The first of those principles is interdependence. All members of an ecological community are interconnected in a vast and intricate network of relationships, the web of life. They derive their essential properties and, in fact, their very existence from their relationships to other things. Interdependence—the mutual dependence of all life processes on one another—is the nature of all ecological relationships. The behavior of every living member of the ecosystem depends on the behavior of many others. The success of the whole community depends on the success of its individual members, while the success of each member depends on the success of the community as a whole.

Understanding ecological interdependence means understanding relationships. It requires the shifts of perception that are characteristic of systems thinking—from the parts to the whole, from objects to relationships, from contents to patterns. A sustainable human community is aware of the multiple relationships among its members. Nourishing the community means nourishing those relationships.

The fact that the basic pattern of life is a network pattern means that the relationships among the members of an ecological community are nonlinear, involving multiple feedback loops. Linear chains of cause and effect exist very rarely in ecosystems. Thus a disturbance will not be limited to a single effect but is likely to spread out in ever-widening patterns. It may even be amplified by interdependent feedback loops, which may completely obscure the original source of the disturbance.

The cyclical nature of ecological processes is an important principle of ecology. The ecosystem's feedback loops are the pathways along which nutrients are continually recycled. Being open systems, all organisms in an ecosystem produce wastes, but what is waste for one species is food for another, so that the ecosystem as a whole remains without waste. Communities of organisms have evolved in this way over billions of years, continually using and recycling the same molecules of minerals, water, and air.

The lesson for human communities here is obvious. A major clash between economics and ecology derives from the fact that nature is cyclical, whereas our industrial systems are linear. Our businesses take resources, transform them into products plus waste, and sell the products to consumers, who discard more waste when they have consumed the products. Sustainable patterns of production and consumption need to be cyclical, imitating the cyclical processes in nature. To achieve such cyclical patterns we need to fundamentally redesign our businesses and our economy.[3]

Ecosystems differ from individual organisms in that they are largely (but not completely) closed systems with respect to the flow of matter, while being open with respect to the flow of energy. The primary source for that flow of energy is the sun. Solar energy, transformed into chemical energy by the photosynthesis of green plants, drives most ecological cycles.

The implications for maintaining sustainable human communities are again obvious. Solar energy in its many forms—sunlight for solar heating and photovoltaic electricity, wind and hydropower, biomass, and so on—is the only kind of energy that is renewable, economically efficient, and environmentally benign. By disregarding this ecological fact, our political and corporate leaders again and again endanger the health and well-being of millions around the world. The 1991 war in the Persian Gulf, for example, which killed hundreds of thousands, impoverished millions, and caused unprecedented environmental disasters, had its roots to a large extent in the misguided energy policies of the Reagan and Bush administrations.

To describe solar energy as economically efficient assumes that the costs of energy production are counted honestly. This is not the case in most of today's market economies. The so-called free market does not provide consumers with proper information, because the social and environmental costs of production are not part of current economic models.[4] These costs are labeled "external" variables by corporate and government economists, because they do not fit into their theoretical framework.

Corporate economists treat as free commodities not only the air, water, and soil, but also the delicate web of social relations, which is severely affected by continuing economic expansion. Private profits are being made at public costs in the deterioration of the environment and the general quality of life, and at the expense of future generations. The marketplace simply gives us the wrong information. There is a lack of feedback, and basic ecological literacy tells us that such a system is not sustainable.

One of the most effective ways to change the situation would be an ecological tax reform. Such a tax reform would be strictly revenue neutral, shifting the tax burden from income taxes to "eco-taxes." This means that taxes would be added to existing products, forms of energy, services, and materials, so that prices would better reflect the true costs.[5] In order to be successful, an ecological tax reform needs to be a slow and long-term process to give new technologies and consumption patterns sufficient time to adapt, and the eco-taxes need to be applied predictably to encourage industrial innovation.

Such a long-term and slow ecological tax reform would gradually drive wasteful and harmful technologies and consumption patterns out of the market. As energy prices go up, with corresponding income tax reductions to offset the increase, people will increasingly switch from cars to bicycles, use public transportation, and carpool on their way to work. As taxes on petrochemicals and fuel go up, again with offsetting reductions in income taxes, organic farming will become not only the healthiest but also the cheapest means of producing food.

Eco-taxes are now under serious discussion in several European countries and are likely to be introduced in all countries sooner or later. To remain competitive under such a new system, managers and entrepreneurs will need to become ecologically literate. In particular, detailed knowledge of the flow of energy and matter through a company will be essential, and this is why the newly developed practice of "eco-auditing" will be of paramount importance.[6] An eco-audit is concerned with the environmental consequences of the flows of material, energy, and people through a company and therefore with the true costs of production.

Partnership is an essential characteristic of sustainable communities. The cyclical exchanges of energy and resources in an ecosystem are sustained by pervasive cooperation. Indeed, we have seen that since the creation of the first nucleated cells over two billion years ago, life on Earth has proceeded through ever more intricate arrangements of cooperation and coevolution. Partnership—the tendency to associate, establish links, live inside one another, and cooperate—is one of the hallmarks of life.

In human communities partnership means democracy and personal empowerment, because each member of the community plays an important role. Combining the principle of partnership with the dynamic of change and development, we may also use the term "coevolution" metaphorically in human communities. As a partnership proceeds, each partner better understands the needs of the other. In a true, committed partnership both partners learn and change—they coevolve. Here again we notice the basic tension between the challenge of ecological sustainability and the way in which our present societies are structured, between economics and ecology. Economics emphasizes competition, expansion, and domination; ecology emphasizes cooperation, conservation, and partnership.

The principles of ecology mentioned so far—interdependence, the cyclical flow of resources, cooperation, and partnership—are all different aspects of the same pattern of organization. This is how ecosystems organize themselves to maximize sustainability.

Once we have understood this pattern, we can ask more detailed questions. For example, what is the resilience of these ecological communities? How do they react to outside disturbances? These questions lead us to two further principles of ecology—flexibility and diversity—that enable ecosystems to survive disturbances and adapt to changing conditions.

The flexibility of an ecosystem is a consequence of its multiple feedback loops, which tend to bring the system back into balance whenever there is a deviation from the norm, due to changing environmental conditions. For example, if an unusually warm summer results in increased growth of algae in a lake, some species of fish feeding on these algae may flourish and breed more, so that their numbers increase and they begin to deplete the algae. Once their major source of food is reduced, the fish will begin to die out. As the fish population drops, the algae will recover and expand again. In this way the original disturbance generates a fluctuation around a feedback loop, which eventually brings the fish/algae system back into balance.

Disturbances of that kind happen all the time, because things in the environment change all the time, and thus the net effect is continual fluctuation. All the variables we can observe in an ecosystem—population densities, availability of nutrients, weather patterns, and so forth—always fluctuate. This is how ecosystems maintain themselves in a flexible state, ready to adapt to changing conditions. The web of life is a flexible, ever-fluctuating network. The more variables are kept fluctuating, the more dynamic is the system; the greater is its flexibility; and the greater is its ability to adapt to changing conditions.

All ecological fluctuations take place between tolerance limits. There is always the danger that the whole system will collapse when a fluctuation goes beyond those limits and the system can no longer compensate for it. The same is true of human communities. Lack of flexibility manifests itself as stress. In particular, stress will occur when one or more variables of the system are pushed to their extreme values, which induces increased rigidity throughout the system. Temporary stress is an essential aspect of life, but prolonged stress is harmful and destructive to the system. These considerations lead to the important realization that managing a social system—a company, a city, or an economy—means finding the *optimal* values for the system's variables. If one tries to maximize any single variable instead of optimizing it, this will invariably lead to the destruction of the system as a whole.

The principle of flexibility also suggests a corresponding strategy of conflict resolution. In every community there will invariably be contradictions and conflicts, which cannot be resolved in favor of one or the other side. For example, the community will need stability *and* change, order *and* freedom, tradition *and* innovation. Rather than by rigid decisions, these unavoidable conflicts are much better resolved by establishing a dynamic balance. Ecological literacy includes the knowledge that both sides of a conflict can be important, depending on the context, and that the contradictions within a community are signs of its diversity and vitality and thus contribute to the system's viability.

In ecosystems the role of diversity is closely connected with the system's network structure. A diverse ecosystem will also be resilient, because it contains many species with overlapping ecological functions that can partially replace one another. When a particular species is destroyed by a severe disturbance so that a link in the network is broken, a diverse community will be able to survive and reorganize itself, because other links in the network can at least partially fulfill the function of the destroyed species. In other words, the more complex the network is, the more complex its pattern of interconnections, the more resilient it will be.

In ecosystems the complexity of the network is a consequence of its biodiversity, and thus a diverse ecological community is a resilient community. In human communities ethnic and cultural diversity may play the same role. Diversity means many different relationships, many different approaches to the same problem. A diverse community is a resilient community, capable of adapting to changing situations.

However, diversity is a strategic advantage only if there is a truly vibrant community, sustained by a web of relationships. If the community is fragmented into isolated groups and individuals, diversity can easily become a source of prejudice and friction. But if the community is aware of the interdependence of all its members, diversity will enrich all the relationships and thus enrich the community as a whole, as well as each individual member. In such a community information and ideas flow freely through the entire network, and the diversity of interpretations and learning styles—even the diversity of mistakes—will enrich the entire community.

These, then, are some of the basic principles of ecology—interdependence, recycling, partnership, flexibility, diversity, and, as a consequence of all those, sustainability. As our century comes to a close and we go toward the beginning of a new millennium, the survival of humanity will depend on our ecological literacy, on our ability to understand these principles of ecology and live accordingly.

Appendix: Bateson Revisited

In this appendix I shall examine Bateson's six criteria of mental process and compare them to the Santiago theory of cognition.[1]

1. *A mind is an aggregate of interacting parts or components.*

This criterion is implicit in the concept of an autopoietic network, which is a network of interacting components.

2. *The interaction between parts of mind is triggered by difference.*

According to the Santiago theory, a living organism brings forth a world by making distinctions. Cognition results from a pattern of distinctions, and distinctions are perceptions of difference. For example, a bacterium, as mentioned on page 268, perceives differences in chemical concentration and temperature.

Thus both Maturana and Bateson emphasize difference, but whereas for Maturana the particular characteristics of a difference are part of the world that is brought forth in the process of cognition, Bateson, as Dell points out, treats differences as objective features of the world. This is apparent in the way Bateson introduces his notion of difference in *Mind and Nature:*

> All receipt of information is necessarily the receipt of news of *difference,* and all perception of difference is limited by threshold. Differences that are too slight or too slowly presented are not perceivable.[2]

In Bateson's view, then, differences are objective features of the world, but not all differences are perceivable. He calls those that are not perceived "potential differences" and those that are "effective differences." The effective differences, Bateson explains, become items of information, and he offers this definition: "Information consists of differences that make a difference."[3]

With this definition of information as effective differences, Bateson comes very close to Maturana's notion that perturbations from the environment trigger structural changes in a living organism. Bateson also emphasizes that different organisms perceive different kinds of differences and that there is no objective information or objective knowledge. However, he holds on to the view that objectivity exists "out there" in the physical world, even though we cannot know it. The idea of differences as objective features of the world becomes more explicit in Bateson's last two criteria of mental process.

3. *Mental process requires collateral energy.*

With this criterion Bateson emphasizes the distinction between the ways living and nonliving systems interact with their environments. Like Maturana, he clearly distinguishes between the reaction of a material object and the response of a living organism. But whereas Maturana describes the autonomy of the organism's response in terms of structural coupling and nonlinear patterns of organization, Bateson characterizes it in terms of energy. "When I kick a stone," he argues, "I give energy to the stone, and it moves with that energy. . . . When I kick a dog, it responds with energy [it received] from [its] metabolism."[4]

However, Bateson was well aware that nonlinear patterns of organization are a principal characteristic of living systems, as his next criterion shows.

4. *Mental process requires circular (or more complex) chains of determination.*

The characterization of living systems in terms of nonlinear patterns of causality was the key that led Maturana to the concept of autopoiesis, and nonlinear causality is also a key ingredient in Prigogine's theory of dissipative structures.

Bateson's first four criteria of mental process, then, are all implicit in the Santiago theory of cognition. In his last two criteria, however, the crucial difference between Bateson's and Maturana's views of cognition becomes apparent.

5. *In mental process, the effects of difference are to be regarded as transforms (that is, coded versions) of events that preceded them.*

Here Bateson explicitly assumes the existence of an independent world, consisting of objective features such as objects, events, and differences. This independently existing outer reality is then "transformed," or "encoded," into an inner reality. In other words, Bateson adheres to the idea that cognition involves mental representations of an objective world.

Bateson's last criterion elaborates the "representationist" position further.

6. *The description and classification of these processes of transformation disclose a hierarchy of logical types immanent in the phenomena.*

To explain this criterion Bateson uses the example of two organisms communicating with each other. Following the computational model of cognition, he describes communication in terms of messages—that is, objective physical signals, such as sounds—that are sent from one organism to the other and then encoded (that is, transformed into mental representations).

In such communications, Bateson argues, the exchanged information will consist not only of messages, but also of messages about coding, which constitute a different class of information. They are messages about messages, or "meta-messages," which Bateson characterizes as being of a different "logical type," borrowing this term from the philosophers Bertrand Russell and Alfred North Whitehead. This proposition then naturally leads Bateson to postulate "messages about meta-messages," and so on—in other words, a "hierarchy of logical types." The existence of such a hierarchy of logical types is Bateson's last criterion of mental process.

The Santiago theory, too, provides a description of communication among living organisms. In Maturana's view communication does not involve any exchange of messages or information, but it does include "communication about communication" and thus what Bateson calls a hierarchy of logical types. However, according to Maturana, such a hierarchy emerges with human language and self-awareness and is not characteristic of the general phenomenon of cognition.[5] With human language arise abstract thinking, concepts, symbols, mental representations, self-awareness, and all the other qualities of consciousness. In Maturana's view Bateson's codes, "transforms," and logical types—his last two criteria—are characteristics not of cognition in general, but of human consciousness.

During the last years of his life Bateson struggled to find additional criteria that would apply to consciousness. Although he suspected that "the phenomenon is somehow related to the business of logical types,"[6] he failed to recognize his last two criteria as criteria of consciousness, rather than mental process. I believe that this error may have prevented Bateson from gaining further insights into the nature of the human mind.

Notes

PREFACE

1. Quoted in Judson (1979), pp. 209, 220.

CHAPTER 1

1. One of the best sources is *State of the World,* a series of annual reports published by the Worldwatch Institute in Washington, D.C. Other excellent accounts can be found in Hawken (1993) and Gore (1992).
2. Brown (1981).
3. See Capra (1975).
4. Kuhn (1962).
5. See Capra (1982).
6. Capra (1986).
7. See Devall and Sessions (1985).
8. See Capra and Steindl-Rast (1991).
9. Arne Naess, quoted in Devall and Sessions (1985), p. 74.
10. See Merchant (1994), Fox (1989).
11. See Bookchin (1981).
12. Eisler (1987).
13. See Merchant (1980).
14. See Spretnak (1978, 1993).
15. See Capra (1982), p. 43.

16. See p. 35 below.
17. Arne Naess, quoted in Fox (1990), p. 217.
18. See Fox (1990), pp. 246–47.
19. Macy (1991).
20. Fox (1990).
21. Roszak (1992).
22. Quoted in Capra (1982), p. 55.

CHAPTER 2

1. See pp. 132–33 below.
2. Bateson (1972), p. 449.
3. See Windelband (1901), pp. 139ff.
4. See Capra (1982), pp. 53ff.
5. R. D. Laing, quoted in Capra (1988), p. 133.
6. See Capra (1982), pp. 107–8.
7. Blake (1802).
8. See Capra (1983), p. 6.
9. See Haraway (1976), pp. 40–42.
10. See Windelband (1901), p. 565.
11. See Webster and Goodwin (1982).
12. Kant (1790, 1987 edition), p. 253.
13. See p. 83 below.
14. See Spretnak (1981), pp. 30ff.
15. See Gimbutas (1982).
16. See pp. 85ff. below.
17. See Sachs (1995).
18. See Webster and Goodwin (1982).
19. See Capra (1982), pp. 108ff.
20. See Haraway (1976), pp. 22ff.
21. Koestler (1967).
22. See Driesch (1908), pp. 76ff.
23. Sheldrake (1981).
24. See Haraway (1976), pp. 33ff.
25. See Lilienfeld (1978), p. 14.
26. I am grateful to Heinz von Foerster for this observation.
27. See Haraway (1976), pp. 131, 194.
28. Quoted ibid., p. 139.
29. See Checkland (1981), p. 78.
30. See Haraway (1976), pp. 147ff.

31. Quoted in Capra (1975), p. 264.

32. Quoted ibid., p. 139.

33. Unfortunately Heisenberg's British and American publishers did not realize the significance of this title and retitled the book *Physics and Beyond;* see Heisenberg (1971).

34. See Lilienfeld (1978), pp. 227ff.

35. Christian von Ehrenfels, "Über 'Gestaltqualitäten,' " 1890; reprinted in Weinhandl (1960).

36. See Capra (1982), p. 427.

37. See Heims (1991), p. 209.

38. Ernst Haeckel, quoted in Maren-Grisebach (1982), p. 30.

39. Uexküll (1909).

40. See Ricklefs (1990), pp. 174ff.

41. See Lincoln et al. (1982).

42. Vernadsky (1926); see also Marhulis and Sagan (1995), pp. 44ff.

43. See pp. 100ff. below.

44. See Thomas (1975), pp. 26ff., 102ff.

45. Ibid.

46. See Burns et al. (1991).

47. Patten (1991).

CHAPTER 3

1. I owe this insight to my brother, Bernt Capra, who was trained as an architect.

2. Quoted in Capra (1988), p. 66.

3. Quoted ibid.

4. Quoted ibid.

5. See ibid., pp. 50ff.

6. Quoted in Capra (1975), p. 126.

7. Quoted in Capra (1982), p. 101.

8. Odum (1953).

9. Whitehead (1929).

10. Cannon (1932).

11. I am grateful to Vladimir Maikov and his colleagues at the Russian Academy of Sciences for introducing me to Bogdanov's work.

12. Quoted in Gorelik (1975).

13. For a detailed summary of tektology, see Gorelik (1975).

14. See pp. 51ff. below.

15. See p. 158 below.

16. See pp. 86ff. below.
17. See p. 134 below.
18. See pp. 56ff. below.
19. See pp. 112ff. below.
20. See Mattessich (1983–84).
21. Quoted in Gorelik (1975).
22. See Bertalanffy (1940) for his first discussion of open systems, published in German, and Bertalanffy (1950) for his first essay on open systems in English, reprinted in Emery (1969).
23. See pp. 75ff. below.
24. See Davidson (1983); see also Lilienfeld (1978), pp. 16–26, for a short review of Bertalanffy's work.
25. Bertalanffy (1968), p. 37.
26. See Capra (1982), pp. 72ff.
27. The "first law of thermodynamics" is the law of the conservation of energy.
28. The term represents a combination of "energy" and *tropos,* the Greek word for transformation, or evolution.
29. Bertalanffy (1968), p. 121.
30. See pp. 185ff. below.
31. See pp. 86ff. below.
32. Bertalanffy (1968), p. 84.
33. Ibid., pp. 80–81.

CHAPTER 4

1. Wiener (1948). The phrase appears in the subtitle of the book.
2. Wiener (1950), p. 96.
3. See Heims (1991).
4. See Varela et al. (1991), p. 38.
5. See Heims (1991).
6. See Heims (1980).
7. Quoted ibid., p. 208.
8. See Capra (1988), pp. 73ff.
9. See pp. 172ff. below.
10. See Heims (1991), pp. 19ff.
11. Wiener (1950), p. 24.
12. See Richardson (1991), pp. 17ff.
13. Cited ibid., p. 94.
14. Cannon (1932).

15. See Richardson (1991), pp. 5–7.

16. In slightly more technical language, the "+" and "–" labels are called "polarities," and the rule says that the polarity of a feedback loop is the product of the polarities of its causal links.

17. Wiener (1948), p. 24.

18. See Richardson (1991), pp. 59ff.

19. See ibid., pp. 79ff.

20. Maruyama (1963).

21. See Richardson (1991), p. 204.

22. See p. 158 below.

23. Heinz von Foerster, private communication, January 1994.

24. Ashby (1952), p. 9.

25. Wiener (1950), p. 32.

26. Ashby (1956), p. 4.

27. See Varela et al. (1992), pp. 39ff.

28. Quoted in Weizenbaum (1976), p. 138.

29. See ibid., pp. 23ff.

30. Quoted in Capra (1982), p. 47.

31. See p. 274 below.

32. See p. 284 below.

33. Weizenbaum (1976), pp. 8, 226.

34. Wiener (1948), p. 38.

35. Wiener (1950), p. 162.

36. Postman (1992), Mander (1991).

37. Postman (1992), p. 19.

38. See Sloan (1985), Kane (1993), Bowers (1993), Roszak (1994).

39. Roszak (1994), pp. 87ff.

40. Bowers (1993), pp. 17ff.

41. See Douglas D. Noble, "The Regime of Technology in Education," in Kane (1993).

42. See Varela et al. (1992), pp. 85ff.

CHAPTER 5

1. See Checkland (1981), pp. 123ff.

2. See ibid., p. 129.

3. See Dickson (1971).

4. Quoted in Checkland (1981), p. 137.

5. See ibid.

6. See Richardson (1992), pp. 149ff. and pp. 170ff.

7. Ulrich (1984).
8. See Königswieser and Lutz (1992).
9. See Capra (1982), pp. 116ff.
10. Lilienfeld (1978), pp. 191–92.
11. See pp. 122–23 below.
12. See pp. 18–19 above.
13. See p. 36 above.
14. See pp. 162ff. below.
15. See Varela et al. (1992), p. 94.
16. See pp. 56ff. above.
17. McCulloch and Pitts (1943).
18. See, e.g., Ashby (1947).
19. See Yovits and Cameron (1959); Foerster and Zopf (1962); and Yovits, Jacobi, and Goldstein (1962).
20. The mathematical definition for redundancy is $R = 1 - H/H\text{max}$, where H is the entropy of the system at a given time and Hmax is the maximum entropy possible for that system.
21. For a detailed review of the history of these research projects, see Paslack (1991).
22. Quoted ibid., p. 97n.
23. See Prigogine and Stengers (1984), p. 142.
24. See Laszlo (1987), p. 29.
25. See Prigogine and Stengers (1984), pp. 146ff.
26. Ibid., p. 143.
27. Prigogine (1967).
28. Prigogine and Glansdorff (1971).
29. Quoted in Paslack (1991), p. 105.
30. See Graham (1987).
31. See Paslack (1991), pp. 106–7.
32. Quoted ibid., p. 108; see also Haken (1987).
33. Reprinted in Haken (1983).
34. Graham (1987).
35. Quoted in Paslack (1991), p. 111.
36. Eigen (1971).
37. See Prigogine and Stengers (1984), pp. 133ff.; see also Laszlo (1987), pp. 31ff.
38. See Laszlo (1987), pp. 34–35.
39. Quoted in Paslack (1991), p. 112.
40. Humberto Maturana in Maturana and Varela (1980), p. xii.
41. Maturana (1970).

42. Quoted in Paslack (1991), p. 156.
43. Maturana (1970).
44. Quoted in Paslack (1991), p. 155.
45. Maturana (1970); see pp. 162ff. below for more details and examples.
46. See pp. 264ff. below.
47. Humberto Maturana in Maturana and Varela (1980), p. xvii.
48. Maturana and Varela (1972).
49. Varela, Maturana, and Uribe (1974).
50. Maturana and Varela (1980), p. 75.
51. See p. 18 and p. 65 above.
52. Maturana and Varela (1980), p. 82.
53. See Capra (1985).
54. Geoffrey Chew, quoted in Capra (1975), p. 296.
55. See below, pp. 158ff.
56. See pp. 22–23 and 33 above.
57. See Kelley (1988).
58. See Lovelock (1979), pp. 1ff.
59. Lovelock (1991), pp. 21–22.
60. Ibid., p. 12.
61. See Lovelock (1979), p. 11.
62. Lovelock (1972).
63. Margulis (1989).
64. See Lovelock (1991), pp. 108–11; see also Harding (1994).
65. Margulis (1989).
66. See Lovelock and Margulis (1974).
67. Lovelock (1991), p. 11.
68. See pp. 24ff. above.
69. See pp. 220, 232 below.
70. See Lovelock (1991), p. 62.
71. See ibid., pp. 62ff.; see also Harding (1994).
72. Harding (1994).
73. See Lovelock (1991), pp. 70–72.
74. See Schneider and Boston (1991).
75. Jantsch (1980).

CHAPTER 6

1. Quoted in Capra (1982), p. 55.
2. Quoted in Capra (1982), p. 63.
3. Stewart (1989), p. 38.

4. Quoted ibid., p. 51.

5. To be precise, the pressure is the force divided by the area the gas is pushing against.

6. A technical point should perhaps be made here. Mathematicians distinguish between dependent and independent variables. In the function $y = f(x)$, y is the dependent variable and x the independent variable. Differential equations are called "linear" when all *dependent* variables appear in the first power, while independent variables may appear in higher powers, and "nonlinear" when *dependent* variables appear in higher powers. See also pp. 115–16 above.

7. See Stewart (1989), p. 83.

8. See Briggs and Peat (1989), pp. 52ff.

9. See Stewart (1989), pp. 155ff.

10. See Stewart (1989), pp. 95–96.

11. See p. 121 above.

12. Quoted in Stuart (1989), p. 71.

13. Ibid., p. 72. See pp. 129ff. below for a detailed discussion of strange attractors.

14. See Capra (1982), pp. 75ff.

15. See Prigogine and Stengers (1984), p. 247.

16. See Mosekilde et al. (1988).

17. See Gleick (1987), pp. 11ff.

18. Quoted in Gleick (1987), p. 18.

19. See Stewart (1989), pp. 106ff.

20. See pp. 86ff. above.

21. See Briggs and Peat (1989), pp. 84ff.

22. Abraham and Shaw (1982–88).

23. Mandelbrot (1983).

24. See Peitgen et al. (1990). This videotape, which contains stunning computer animation and captivating interviews with Benoît Mandelbrot and Edward Lorenz, is one of the best introductions to fractal geometry.

25. See ibid.

26. Ibid.

27. See Mandelbrot (1983), pp. 34ff.

28. See Dantzig (1954), pp. 181ff.

29. Quoted in Dantzig (1954), p. 204.

30. Quoted ibid., p. 189.

31. Quoted ibid., p. 190.

32. See Gleick (1987), pp. 221ff.

33. For real numbers it is easy to see that any number greater than 1 will

keep increasing when it is squared repeatedly, while any number smaller than 1 will keep decreasing. Adding a constant at every step of the iteration before squaring again adds more variety, and for complex numbers the whole situation becomes even more complicated.

34. Quoted in Gleick (1987), pp. 221–22.

35. See Peitgen et al. (1990).

36. See Peitgen et al. (1990).

37. See Peitgen and Richter (1986).

38. See Grof (1976).

39. Quoted in Peitgen et al. (1990).

40. Quoted in Gleick (1987), p. 52.

CHAPTER 7

1. Maturana and Varela (1987), p. 47. Instead of "pattern of organization" the authors simply use the term "organization."

2. See pp. 18–19 above.

3. See pp. 95ff. above.

4. See pp. 86ff. above.

5. See above, pp. 86–88.

6. See above, pp. 82–83.

7. Maturana and Varela (1980), p. 49.

8. See Capra (1982), p. 119.

9. See p. 243 below.

10. To do so, the enzymes use the other, complementary DNA strand as a template for the section to be replaced. The double strandedness of DNA is thus essential for these repair processes.

11. I am grateful to William Holloway for research assistance on vortex phenomena.

12. Technically speaking, this effect is a consequence of the conservation of angular momentum.

13. See pp. 136–37 above.

14. See pp. 190–92 below.

15. See p. 55 above.

16. Bateson's first published discussions of these criteria, initially called "mental characteristics," can be found in two essays, "The Cybernetics of 'Self': A Theory of Alcoholism" and "Pathologies of Epistemology," both reprinted in Bateson (1972). For a more comprehensive discussion, see Bateson (1979), pp. 89ff. See appendix, pp. 305ff. below, for a detailed discussion of Bateson's criteria of mental process.

17. See Bateson (1972), p. 478.
18. See p. 96 above.
19. Bateson (1979), p. 8.
20. Quoted in Capra (1988), p. 88.
21. See pp. 95–96 above.
22. See pp. 264ff. below.
23. Revonsuo and Kamppinen (1994), p. 5.
24. See pp. 282ff. below.

CHAPTER 8

1. See p. 48 above.
2. Odum (1953).
3. Prigogine and Stengers (1984), p. 156.
4. See pp. 86ff. above.
5. Prigogine and Stengers (1984), pp. 22–23.
6. Ibid., pp. 143–44.
7. See pp. 112ff. above.
8. Prigogine and Stengers (1984), p. 140.
9. See p. 126 above.
10. Prigogine (1989).
11. Quoted in Capra (1975), p. 45.
12. I have used the general term "catalytic loops" to refer to many complex nonlinear relationships between catalysts, including autocatalysis, cross-catalysis, and autoinhibition. For more details, see Prigogine and Stengers (1984), p. 153.
13. Prigogine and Stengers (1984), p. 292.
14. See pp. 13 above.
15. See p. 47 above.
16. Prigogine and Stengers (1984), p. 129.
17. See pp. 121–22 above.
18. See Prigogine and Stengers (1984), pp. 123–24.
19. If N is the total number of particles, and if N_1 particles are on one side and N_2 on the other, the number of different possibilities is given by $P = N! / N_1!\ N_2!$, where $N!$ is a shorthand notation for $1 \times 2 \times 3 \ldots \times N$.
20. Prigogine (1989).
21. See Briggs and Peat (1989), pp. 45ff.
22. See Prigogine and Stengers (1984), pp. 144ff.
23. See Prigogine (1980), pp. 104ff.
24. Goodwin (1994), pp. 89ff.

25. See p. 220 below.
26. Prigogine and Stengers (1984), p. 176.
27. Prigogine (1989).

CHAPTER 9

1. See p. 88 above.
2. See p. 97 above.
3. See pp. 107ff. above.
4. See p. 83 above.
5. Von Neumann (1966).
6. See Gardner (1971).
7. In each three-by-three area there is a center cell surrounded by eight neighbors. If three neighboring cells are black, the center becomes black at the next step ("birth"); if two neighbors are black, the center cell is left unchanged ("survival"); in all other cases the center becomes white ("death").
8. See Gardner (1970).
9. For an excellent account of the history and applications of cellular automata, see Farmer, Toffoli, and Wolfram (1984), especially the preface by Stephen Wolfram. For a more recent and more technical collection of papers, see Gutowitz (1991).
10. Varela, Maturana, and Uribe (1974).
11. These movements and interactions can be expressed formally as mathematical transition rules that apply simultaneously to all cells.
12. Some of the corresponding mathematical probabilities serve as variable parameters of the model.
13. The disintegration probability must be less than 0.01 per time step to achieve any viable structure at all, and the boundary must contain at least ten links; see Varela, Maturana, and Uribe (1974) for further details.
14. See Kauffman (1993), pp. 182ff.; see also Kauffman (1991) for a short summary.
15. See pp. 127ff. above. Note, however, that since the values of the binary variables vary discontinuously, their phase space, too, is discontinuous.
16. See Kauffman (1993), p. 183.
17. See ibid., p. 191.
18. Ibid., pp. 441ff.
19. See pp. 66ff. above.
20. Varela et al. (1992), p. 188.
21. Kauffman (1991).

22. See Kauffman (1993), p. 479.
23. Kauffman (1991).
24. See Luisi and Varela (1989), Bachmann et al. (1990), Walde et al. (1994).
25. See Fleischaker (1990).
26. See Fleischaker (1992) for a recent debate on many of the issues discussed in the following pages; see also Mingers (1995).
27. Maturana and Varela (1987), p. 89.
28. See pp. 286ff. below.
29. Maturana and Varela (1987), p. 199.
30. See Fleischaker (1992); Mingers (1995), pp. 119ff.
31. Mingers (1995), p. 127.
32. See Fleischaker (1992), pp. 131–41; Mingers (1995), pp. 125–26.
33. Maturana (1988); see also pp. 290–91 below.
34. Varela (1981).
35. Luhmann (1990).
36. See p. 104 above.
37. See pp. 100ff. above.
38. Lovelock (1991), pp. 31ff.
39. See p. 208 above.
40. See p. 93 above.
41. See Lovelock (1991), pp. 135–36.
42. Harding (1994).
43. See Margulis and Sagan (1986), p. 66.
44. Margulis (1993); Margulis and Sagan (1986).
45. See pp. 236ff. below.
46. Margulis and Sagan (1986), pp. 14, 21.
47. Ibid., p. 271.
48. Quoted in Capra (1975), p. 183.
49. See pp. 233ff. below.
50. See Lovelock (1991), p. 127.
51. See Maturana and Varela (1987), pp. 75ff.
52. Ibid., p. 95.

CHAPTER 10

1. See Capra (1982), pp. 116ff.
2. Quoted ibid., p. 114.
3. Margulis (1995).
4. See pp. 228ff. below.

5. See pp. 204–5 above.
6. See Gould (1994).
7. Kauffman (1993), pp. 173, 408, 644.
8. See Jantsch (1980) and Laszlo (1987) for early attempts of a synthesis of some of those elements.
9. Lovelock (1991), p. 99.
10. See Margulis and Sagan (1986), pp. 15ff.
11. See Capra (1982), pp. 118–19.
12. See Margulis and Sagan (1986), p. 75.
13. Ibid., p. 16.
14. Ibid., p. 89.
15. See ibid.
16. See ibid.
17. Margulis (1995).
18. See p. 164 above.
19. Margulis and Sagan (1986), p. 17.
20. Ibid., p. 15.
21. Margulis and Sagan (1986); see also Margulis and Sagan (1995) and Calder (1983).
22. Margulis and Sagan (1986), p. 51.
23. See pp. 93–94 above; see also Kauffman (1993), pp. 287ff.
24. See p. 208 above.
25. Margulis and Sagan (1986), p. 64.
26. See p. 164 above.
27. Margulis and Sagan (1986), p. 78.
28. See Lovelock (1991), pp. 80ff.
29. See Margulis (1993), pp. 160ff.
30. See pp. 166–67 above.
31. Margulis and Sagan (1986), p. 93.
32. Ibid., p. 191.
33. Ibid., p. 103.
34. Ibid., p. 109.
35. See Lovelock (1991), pp. 113ff.
36. See pp. 162ff. above.
37. See pp. 230ff. above.
38. Margulis and Sagan (1986), p. 119.
39. See p. 165 above.
40. See Margulis and Sagan (1986), p. 133.
41. See Thomas (1975), pp. 141ff.
42. Margulis and Sagan (1986), pp. 155ff.

43. See Margulis, Schwartz, and Dolan (1994).
44. Margulis and Sagan (1986), p. 174.
45. Ibid., p. 73.
46. See Margulis and Sagan (1995), pp. 140ff.
47. Margulis and Sagan (1986), p. 214.
48. See ibid., pp. 208ff.
49. Ibid., p. 210.
50. Brower (1995), p. 18.
51. See *New York Times,* June 8, 1995; Chauvet et al. (1995).
52. Margulis and Sagan (1986), pp. 223–24.

CHAPTER 11

1. See pp. 174–75 above.
2. See Windelband (1901), pp. 232–33.
3. See pp. 173ff. above.
4. See Varela et al. (1991), pp. 4ff.
5. See pp. 66ff. above.
6. See Varela et al. (1991), pp. 8, 41.
7. Ibid., pp. 93–94.
8. See Gluck and Rumelhart (1990).
9. Varela et al. (1991), p. 94.
10. See p. 97 above.
11. See ibid.
12. See pp. 218–20 above.
13. Maturana and Varela (1987), p. 174.
14. See Margulis and Sagan (1995), p. 179.
15. Varela et al. (1991), p. 200.
16. Ibid., p. 177.
17. See pp. 287ff. below.
18. See p. 284 below.
19. See p. 290 below.
20. Varela et al. (1991), p. 135.
21. See p. 290 below.
22. Varela et al. (1991), p. 140.
23. Ibid., p. 101.
24. See p. 173 above.
25. Dell (1985).
26. See appendix, pp. 305ff. below.
27. Winograd and Flores (1991), p. 97.

28. See ibid., pp. 93ff.
29. Ibid., pp. 107ff.
30. Ibid. p. 113.
31. Ibid., pp. 133ff.
32. Ibid., p. 132.
33. Dreyfus and Dreyfus (1986), p. 108.
34. See Varela and Coutinho (1991a).
35. See Varela and Coutinho (1991b).
36. Varela and Coutinho (1991a).
37. Ibid.
38. See Varela and Coutinho (1991b).
39. Francisco Varela, private communication, April 1991.
40. Pert et al. (1985), Pert (1993).
41. Pert (1989).
42. See Pert (1992), Pert (1995).
43. Pert (1989).

CHAPTER 12

1. Maturana (1970), Maturana and Varela (1987), Maturana (1988).
2. Maturana and Varela (1987), pp. 193–94.
3. Humberto Maturana, private communication, 1985.
4. See Maturana and Varela (1987), pp. 212ff.
5. Ibid., p. 215.
6. See appendix, pp. 307–8 below.
7. Maturana and Varela (1987), p. 234.
8. Ibid., p. 245.
9. Ibid., p. 244.
10. See Capra (1982), p. 302.
11. See Capra (1975), p. 88.
12. Varela (1995).
13. See Capra (1982), p. 178.
14. See p. 259 above.
15. See Varela et al. (1991), pp. 217ff.
16. See Capra (1975), pp. 93ff.
17. See Varela et al. (1991), pp. 59ff.
18. Ibid., p. 143.
19. Margulis and Sagan (1995), p. 26.

EPILOGUE

1. See Orr (1992).

2. For applications of the principles of ecology to education, see Capra (1993); for applications to business, see Callenbach et al. (1993), Capra and Pauli (1995); for applications to politics, see Henderson (1995).

3. See Hawken (1993).

4. See ibid., pp. 75ff.; see also Henderson (1995).

5. See Hawken (1993), pp. 177ff.; Daly (1995).

6. See Callenbach et al. (1993).

APPENDIX

1. Bateson (1979), pp. 89ff. See pp. 173ff. above and pp. 273ff. above for the historical and philosophical contexts of Bateson's concept of mental process.

2. Bateson (1979), p. 29.

3. Ibid., p. 99.

4. Ibid., p. 101.

5. See p. 290 above.

6. Bateson (1979), p. 128.

Bibliography

Abraham, Ralph H., and Christopher D. Shaw, *Dynamics: The Geometry of Behavior,* vols. 1–4, Aerial Press, Santa Cruz, Calif., 1982–88.

Ashby, Ross, "Principles of the Self-Organizing Dynamic System," *Journal of General Psychology,* vol. 37, p. 125, 1947.

Ashby, Ross, *Design for a Brain,* John Wiley, New York, 1952.

Ashby, Ross, *Introduction to Cybernetics,* John Wiley, New York, 1956.

Bachmann, Pascale Angelica, Peter Walde, Pier Luigi Luisi, and Jacques Lang, "Self-Replicating Reverse Micelles and Chemical Autopoiesis," *Journal of the American Chemical Society,* 112, 8200–8201, 1990.

Bateson, Gregory, *Steps to an Ecology of Mind,* Ballantine, New York, 1972.

Bateson, Gregory, *Mind and Nature: A Necessary Unity,* Dutton, New York, 1979.

Bergé, P., "Rayleigh-Bénard Convection in High Prandtl Number Fluid," in H. Haken, *Chaos and Order in Nature,* Springer, New York, 1981; pp. 14–24.

Bertalanffy, Ludwig von, "Der Organismus als physikalisches System betrachtet," *Die Naturwissenschaften,* vol. 28, pp. 521–31, 1940.

Bertalanffy, Ludwig von, "The Theory of Open Systems in Physics and Biology," *Science,* vol. 111, pp. 23–29, 1950.

Bertalanffy, Ludwig von, *General System Theory,* Braziller, New York, 1968.

Blake, William, letter to Thomas Butts, 22 November 1802; in Alicia Ostriker (ed.), *William Blake: The Complete Poems,* Penguin, New York, 1977.

BOOKCHIN, MURRAY, *The Ecology of Freedom,* Cheshire Books, Palo Alto, Calif., 1981.

BOWERS, C. A., *Critical Essays on Education, Modernity, and the Recovery of the Ecological Imperative,* Teachers College Press, New York, 1993.

BRIGGS, JOHN, and F. DAVID PEAT, *Turbulent Mirror,* Harper & Row, New York, 1989.

BROWER, DAVID, *Let the Mountains Talk, Let the Rivers Run,* HarperCollins, New York, 1995.

BROWN, LESTER R., *Building a Sustainable Society,* Norton, New York, 1981.

BROWN, LESTER R., et al., *State of the World,* Norton, New York, 1984–94.

BURNS, T. P., B. C. PATTEN, and H. HIGASHI, "Hierarchical Evolution in Ecological Networks," in Higashi, M., and T. P. Burns, *Theoretical Studies of Ecosystems: The Network Perspective,* Cambridge University Press, New York, 1991.

BUTTS, ROBERT, and JAMES BROWN (eds.), *Constructivism and Science,* Kluwer, Dordrecht, The Netherlands, 1989.

CALDER, NIGEL, *Timescale,* Viking, New York, 1983.

CALLENBACH, ERNEST, FRITJOF CAPRA, LENORE GOLDMAN, SANDRA MARBURG, and RÜDIGER LUTZ, *EcoManagement,* Berrett-Koehler, San Francisco, 1993.

CANNON, WALTER B., *The Wisdom of the Body,* Norton, New York, 1932; rev. ed., 1939.

CAPRA, FRITJOF, *The Tao of Physics,* Shambhala, Boston, 1975; 3rd updated ed., 1991.

CAPRA, FRITJOF, *The Turning Point,* Simon & Schuster, New York, 1982.

CAPRA, FRITJOF, *Wendezeit* (German edition of *The Turning Point),* Scherz, 1983.

CAPRA, FRITJOF, "Bootstrap Physics: A Conversation with Geoffrey Chew," in Carleton deTar, Jerry Finkelstein, and Chung-I Tan (eds.), *A Passion for Physics,* World Scientific, Singapore, 1985; pp. 247–86.

CAPRA, FRITJOF, "The Concept of Paradigm and Paradigm Shift," *Re-Vision,* vol. 9, no. 1, p. 3, 1986.

CAPRA, FRITJOF, *Uncommon Wisdom,* Simon & Schuster, New York, 1988.

CAPRA, FRITJOF, and DAVID STEINDL-RAST, with Thomas Matus, *Belonging to the Universe,* Harper & Row, San Francisco, 1991.

CAPRA, FRITJOF (ed.), *Guide to Ecoliteracy,* 1993; available from Center for Ecoliteracy, 2522 San Pablo Ave., Berkeley, Calif. 94702.

CAPRA, FRITJOF, and GUNTER PAULI (eds.), *Steering Business toward Sustainability,* United Nations University Press, Tokyo, 1995.

CHAUVET, JEAN-MARIE, ÉLIETTE BRUNEL DESCHAMPS, and CHRISTIAN HILLAIRE, *La Grotte Chauvet à Vallon-Pont-d'Arc,* Seuil, Paris, 1995.

CHECKLAND, PETER, *Systems Thinking, Systems Practice,* John Wiley, New York, 1981.

DANTZIG, TOBIAS, *Number: The Language of Science,* 4th ed., Macmillan, New York, 1954.

DALY, HERMAN, "Ecological Tax Reform," in Capra and Pauli (1995), pp. 108–24.

DAVIDSON, MARK, *Uncommon Sense: The Life and Thought of Ludwig von Bertalanffy,* Tarcher, Los Angeles, 1983.

DELL, PAUL, "Understanding Maturana and Bateson," *Journal of Marital and Family Therapy,* vol. 11, no. 1, pp. 1–20, 1985.

DEVALL, BILL, and GEORGE SESSIONS, *Deep Ecology,* Peregrine Smith, Salt Lake City, Utah, 1985.

DICKSON, PAUL, *Think Tanks,* Atheneum, New York, 1971.

DREYFUS, HUBERT, and STUART DREYFUS, *Mind over Machine,* Free Press, New York, 1986.

DRIESCH, HANS, *The Science and Philosophy of the Organism,* Aberdeen University Press, Aberdeen, 1908.

EIGEN, MANFRED, "Molecular Self-Organization and the Early Stages of Evolution," *Quarterly Reviews of Biophysics,* 4, 2&3, 149, 1971.

EISLER, RIANE, *The Chalice and the Blade,* Harper & Row, San Francisco, 1987.

EMERY, F. E. (ed.), *Systems Thinking: Selected Readings,* Penguin, New York, 1969.

FARMER, DOYNE, TOMMASO TOFFOLI, and STEPHEN WOLFRAM (eds.), *Cellular Automata,* North-Holland, 1984.

FLEISCHAKER, GAIL RANEY, "Origins of Life: An Operational Definition," *Origins of Life and Evolution of the Biosphere* 20, 127–37, 1990.

FLEISCHAKER, GAIL RANEY (ed.), "Autopoiesis in Systems Analysis: A Debate," *International Journal of General Systems,* vol. 21, no. 2, 1992.

FOERSTER, HEINZ VON, and GEORGE W. ZOPF (eds.), *Principles of Self-Organization,* Pergamon, New York, 1962.

FOX, WARWICK, "The Deep Ecology—Ecofeminism Debate and Its Parallels," *Environmental Ethics* 11, 5–25, 1989.

FOX, WARWICK, *Toward a Transpersonal Ecology,* Shambhala, Boston, 1990.

GARCIA, LINDA, *The Fractal Explorer,* Dynamic Press, Santa Cruz, Calif., 1991.

GARDNER, MARTIN, "The Fantastic Combinations of John Conway's New Solitaire Game 'Life,'" *Scientific American,* 223, 4, pp. 120–23, 1970.

GARDNER, MARTIN, "On Cellular Automata, Self-Reproduction, the Garden of Eden, and the Game 'Life,'" *Scientific American,* 224, 2, pp. 112–17, 1971.

GIMBUTAS, MARIJA, "Women and Culture in Goddess-Oriented Old Europe," in Charlene Spretnak (ed.), *The Politics of Women's Spirituality,* Anchor, New York, 1982.

GLEICK, JAMES, *Chaos,* Penguin, New York, 1987.

GLUCK, MARK, and DAVID RUMELHART, *Neuroscience and Connectionist Theory,* Lawrence Erlbaum, Hillsdale, N.J., 1990.

GOODWIN, BRIAN, *How the Leopard Changed Its Spots,* Scribner, New York, 1994.

GORE, AL, *Earth in the Balance,* Houghton Mifflin, New York, 1992.

GORELIK, GEORGE, "Principal Ideas of Bogdanov's 'Tektology': The Universal Science of Organization," *General Systems,* vol. XX, pp. 3–13, 1975.

GOULD, STEPHEN JAY, "Lucy on the Earth in Stasis," *Natural History,* no. 9, 1994.

GRAHAM, ROBERT, "Contributions of Hermann Haken to Our Understanding of Coherence and Selforganization in Nature," in R. Graham and A. Wunderlin (eds.), *Lasers and Synergetics,* Springer, Berlin, 1987.

GROF, STANISLAV, *Realms of the Human Unconscious,* Dutton, New York, 1976.

GUTOWITZ, HOWARD (ed.), *Cellular Automata: Theory and Experiment,* MIT Press, Cambridge, Mass., 1991.

HAKEN, HERMANN, *Laser Theory,* Springer, Berlin, 1983.

HAKEN, HERMANN, "Synergetics: An Approach to Self-Organization," in F. Eugene Yates (ed.), *Self-Organizing Systems,* Plenum, New York, 1987.

HARAWAY, DONNA JEANNE, *Crystals, Fabrics and Fields: Metaphors of Organicism in Twentieth-Century Developmental Biology,* Yale University Press, New Haven, 1976.

HARDING, STEPHAN, "Gaia Theory," unpublished lecture notes, Schumacher College, Dartington, Devon, England, 1994.

HAWKEN, PAUL, *The Ecology of Commerce,* HarperCollins, New York, 1993.

HEIMS, STEVE J., *John von Neumann and Norbert Wiener,* MIT Press, Cambridge, Mass., 1980.

HEIMS, STEVE J., *The Cybernetics Group,* MIT Press, Cambridge, Mass., 1991.

HEISENBERG, WERNER, *Physics and Beyond,* Harper & Row, New York, 1971.

HENDERSON, HAZEL, *Paradigms in Progress,* Berrett-Koehler, San Francisco, 1995.

JANTSCH, ERICH, *The Self-Organizing Universe,* Pergamon, New York, 1980.

JUDSON, HORACE FREELAND, *The Eighth Day of Creation,* Simon & Schuster, New York, 1979.

KANE, JEFFREY (ed.), *Holistic Education Review,* Special Issue: Technology and Childhood, summer 1993.

KANT, IMMANUEL, *Critique of Judgment,* 1790; trans., Werer S. Pluhar, Hackett, Indianapolis, Ind., 1987.

KAUFFMAN, STUART, "Antichaos and Adaptation," *Scientific American,* August 1991.

KAUFFMAN, STUART, *The Origins of Order,* Oxford University Press, New York, 1993.

KELLEY, KEVIN (ed.), *The Home Planet,* Addison-Wesley, New York, 1988.

KOESTLER, ARTHUR, *The Ghost in the Machine,* Hutchinson, London, 1967.

KÖNIGSWIESER, ROSWITA, and CHRISTIAN LUTZ (eds.), *Das Systemisch Evolutionäre Management,* Orac, Vienna, 1992.

KUHN, THOMAS S., *The Structure of Scientific Revolutions,* University of Chicago Press, Chicago, 1962.

LASZLO, ERVIN, *Evolution,* Shambhala, Boston, 1987.

LILIENFELD, ROBERT, *The Rise of Systems Theory,* John Wiley, New York, 1978.

LINCOLN, R. J., et al., *A Dictionary of Ecology,* Cambridge University Press, New York, 1982.

LORENZ, EDWARD N., "Deterministic Nonperiodic Flow," *Journal of the Atmospheric Sciences,* vol. 20, pp. 130–41, 1963.

LOVELOCK, JAMES, "Gaia As Seen through the Atmosphere," *Atmospheric Environment,* vol. 6, p. 579, 1972.

LOVELOCK, JAMES, *Gaia,* Oxford University Press, New York, 1979.

LOVELOCK, JAMES, *Healing Gaia,* Harmony Books, New York, 1991.

LOVELOCK, JAMES, and LYNN MARGULIS, "Biological Modulation of the Earth's Atmosphere," *Icarus,* vol. 21, 1974.

LUHMANN, NIKLAS, "The Autopoiesis of Social Systems," in Niklas Luhmann, *Essays on Self-Reference,* Columbia University Press, New York, 1990.

LUISI, PIER LUIGI, and FRANCISCO J. VARELA, "Self-Replicating Micelles—A Chemical Version of a Minimal Autopoietic System," *Origins of Life and Evolution of the Biosphere,* 19, 633–43, 1989.

MACY, JOANNA, *World as Lover, World as Self,* Parallax Press, Berkeley, Calif., 1991.

MANDELBROT, BENOÎT, *The Fractal Geometry of Nature,* Freeman, New York, 1983; first French edition published in 1975.

MANDER, JERRY, *In the Absence of the Sacred,* Sierra Club Books, San Francisco, 1991.

MAREN-GRISEBACH, MANON, *Philosophie der Grünen,* Olzog, München, 1982.

MARGULIS, LYNN, "Gaia: The Living Earth," dialogue with Fritjof Capra, *The Elmwood Newsletter,* Berkeley, Calif., vol. 5, no. 2, 1989.

MARGULIS, LYNN, *Symbiosis in Cell Evolution,* 2nd ed., Freeman, San Francisco, 1993.

MARGULIS, LYNN, "Gaia Is a Tough Bitch," in John Brockman, *The Third Culture,* Simon & Schuster, New York, 1995.

MARGULIS, LYNN, and DORION SAGAN, *Microcosmos,* Summit, New York, 1986.

MARGULIS, LYNN, and DORION SAGAN, *What Is Life?* Simon & Schuster, New York, 1995.

MARGULIS, LYNN, KARLENE SCHWARTZ, and MICHAEL DOLAN, *The Illustrated Five Kingdoms,* HarperCollins, New York, 1994.

MARUYAMA, MAGOROH, "The Second Cybernetics," *American Scientist,* vol. 51, pp. 164–79, 1963.

MATTESSICH, RICHARD, "The Systems Approach: Its Variety of Aspects," *General Systems,* vol. 28, pp. 29–40, 1983–84.

MATURANA, HUMBERTO, "Biology of Cognition," published originally in 1970; reprinted in Maturana and Varela (1980).

MATURANA, HUMBERTO, "Reality: The Search for Objectivity or the Quest for a Compelling Argument," *Irish Journal of Psychology,* vol. 9, no. 1, pp. 25–82, 1988.

MATURANA, HUMBERTO, and FRANCISCO VARELA, "Autopoiesis: The Organization of the Living," published originally under the title *De Maquinas y Seres Vivos.* Editorial Universitaria, Santiago, Chile, 1972; reprinted in Maturana and Varela (1980).

MATURANA, HUMBERTO, and FRANCISCO VARELA, *Autopoiesis and Cognition,* D. Reidel, Dordrecht, Holland, 1980.

MATURANA, HUMBERTO, and FRANCISCO VARELA, *The Tree of Knowledge,* Shambhala, Boston, 1987.

MCCULLOCH, WARREN S., and WALTER H. PITTS, "A Logical Calculus of the Ideas Immanent in Nervous Activity," *Bull. of Math. Biophysics,* vol. 5, p. 115, 1943.

MINGERS, JOHN, *Self-Producing Systems,* Plenum, New York, 1995.

MERCHANT, CAROLYN, *The Death of Nature,* Harper & Row, New York, 1980.

MERCHANT, CAROLYN (ed.), *Ecology,* Humanities Press, Atlantic Highlands, N.J., 1994.

MOSEKILDE, ERIK, JAVIER ARACIL, and PETER M. ALLEN, "Instabilities and

Chaos in Nonlinear Dynamic Systems," *System Dynamics Review,* vol. 4, pp. 14–55, 1988.

MOSEKILDE, ERIK, and RASMUS FELDBERG, *Nonlinear Dynamics and Chaos* (in Danish), Polyteknisk Forlag, Lyngby, 1994.

NEUMANN, JOHN VON, *Theory of Self-Reproducing Automata,* edited and completed by Arthur W. Burks, University of Illinois Press, Champaign, Ill., 1966.

ODUM, EUGENE, *Fundamentals of Ecology,* Saunders, Philadelphia, 1953.

ORR, DAVID, *Ecological Literacy,* State University of New York Press, Albany, N.Y., 1992.

PASLACK, RAINER, *Urgeschichte der Selbstorganisation,* Vieweg, Braunschweig, Germany, 1991.

PATTEN, B. C., "Network Ecology," in Higashi, M., and T. P. Burns, *Theoretical Studies of Ecosystems: The Network Perspective,* Cambridge University Press, New York, 1991.

PEITGEN, HEINZ-OTTO, and PETER RICHTER, *The Beauty of Fractals,* Springer, New York, 1986.

PEITGEN, HEINZ-OTTO, HARTMUT JÜRGENS, DIETMAR SAUPE, and C. ZAHLTEN, "Fractals: An Animated Discussion," VHS/color/63 minutes, Freeman, New York, 1990.

PERT, CANDACE, MICHAEL RUFF, RICHARD WEBER, and MILES HERKENHAM, "Neuropeptides and Their Receptors: A Psychosomatic Network," *The Journal of Immunology,* vol. 135, no. 2, pp. 820–26, 1985.

PERT, CANDACE, Presentation at Elmwood Symposium, "Healing Ourselves and Our Society," Boston, December 9, 1989 (unpublished).

PERT, CANDACE, "Peptide T: A New Therapy for AIDS," Elmwood symposium with Candace Pert, San Francisco, November 5, 1992 (unpublished); audiotapes available from Advanced Peptides Inc., 25 East Loop Road, Stony Brook, N.Y. 11790.

PERT, CANDACE, "The Chemical Communicators," interview in Bill Moyers, *Healing and the Mind,* Doubleday, New York, 1993.

PERT, CANDACE, "Neuropeptides, AIDS, and the Science of Mind-Body Healing," interview in *Alternative Therapies,* vol. 1, no. 3, 1995.

POSTMAN, NEIL, *Technopoly,* Knopf, New York, 1992.

PRIGOGINE, ILYA, "Dissipative Structures in Chemical Systems," in Stig Claesson (ed.), *Fast Reactions and Primary Processes in Chemical Kinetics,* Interscience, New York, 1967.

PRIGOGINE, ILYA, *From Being to Becoming,* Freeman, San Francisco, 1980.

PRIGOGINE, ILYA, "The Philosophy of Instability," *Futures,* 21, 4, pp. 396–400 (1989).

PRIGOGINE, ILYA, and PAUL GLANSDORFF, *Thermodynamic Theory of Structure, Stability and Fluctuations,* Wiley, New York, 1971.

PRIGOGINE, ILYA, and ISABELLE STENGERS, *Order out of Chaos,* Bantam, New York, 1984.

REVONSUO, ANTTI, and MATTI KAMPPINEN (eds.), *Consciousness in Philosophy and Cognitive Neuroscience,* Lawrence Erlbaum, Hillsdale, N.J., 1994.

RICHARDSON, GEORGE P., *Feedback Thought in Social Science and Systems Theory,* University of Pennsylvania Press, Philadelphia, 1992.

RICKLEFS, ROBERT E., *Ecology,* 3rd ed., Freeman, New York, 1990.

ROSZAK, THEODORE, *The Voice of the Earth,* Simon & Schuster, New York, 1992.

ROSZAK, THEODORE, *The Cult of Information,* U.C. Press, Berkeley, Calif., 1994.

SACHS, AARON, "Humboldt's Legacy and the Restoration of Science," *World Watch,* March/April 1995.

SCHMIDT, SIEGFRIED (ed.), *Der Diskurs des Radikalen Konstruktivismus,* Suhrkamp, Frankfurt, Germany, 1987.

SCHNEIDER, STEPHEN, and PENELOPE BOSTON (eds.), *Scientists on Gaia,* MIT Press, Cambridge, Mass., 1991.

SHELDRAKE, RUPERT, *A New Science of Life,* Tarcher, Los Angeles, 1981.

SLOAN, DOUGLAS (ed.), *The Computer in Education: A Critical Perspective.* Teachers College Press, New York, 1985.

SPRETNAK, CHARLENE, *Lost Goddesses of Early Greece,* Beacon Press, Boston, 1981.

SPRETNAK, CHARLENE, "An Introduction to Ecofeminism," *Bucknell Review,* Lewisburg, Pennsylvania, 1993.

STEWART, IAN, *Does God Play Dice?* Blackwell, Cambridge, Mass., 1989.

THOMAS, LEWIS, *The Lives of a Cell,* Bantam, New York, 1975.

UEDA, Y., J. S. THOMSEN, J. RASMUSSEN, and E. MOSEKILDE, "Behavior of the Soliton to Duffing's Equation for Large Forcing Amplitudes," *Mathemathical Research* 72, 149–166, 1993.

UEXKÜLL, JAKOB VON, *Umwelt und Innenwelt der Tiere,* Springer, Berlin, 1909.

ULRICH, HANS, *Management,* Haupt, Bern, Switzerland, 1984.

VARELA, FRANCISCO, "Describing the Logic of the Living: The Adequacy and Limitations of the Idea of Autopoiesis," in Milan Zeleny (ed.), *Autopoiesis: A Theory of Living Organization,* North Holland, New York, 1981; pp. 36–48.

VARELA, FRANCISCO, HUMBERTO MATURANA, and RICARDO URIBE, "Autopoiesis: The Organization of Living Systems, Its Characterization and a Model," *BioSystems* 5, 187–96, 1974.

VARELA, FRANCISCO, and ANTONIO COUTINHO, "Immunoknowledge," in J. Brockman (ed.), *Doing Science,* Prentice-Hall, New York, 1991a.

VARELA, FRANCISCO, and ANTONIO COUTINHO, "Second Generation Immune Networks," *Immunology Today,* vol. 12, no. 5, pp. 159–166, 1991b.

VARELA, FRANCISCO, EVAN THOMPSON, and ELEANOR ROSCH, *The Embodied Mind,* MIT Press, Cambridge, Mass., 1991.

VARELA, FRANCISCO, "Resonant Cell Assemblies," *Biological Research,* 28, 81–95, 1995.

VERNADSKY, VLADIMIR, *The Biosphere,* published originally in 1926; reprinted U.S. edition by Synergetic Press, Oracle, Ariz., 1986.

WALDE, PETER, ROGER WICK, MASSIMO FRESTA, ANNAROSA MANGONE, and PIER LUIGI LUISI, "Autopoietic Self-Reproduction of Fatty Acid Vesicles," *Journal of the American Chemical Society,* 116, 11649–54, 1994.

WEBSTER, G., and B. C. GOODWIN, "The Origin of Species: A Structuralist Approach," *Journal of Social and Biological Structures,* vol. 5, pp. 15–47, 1982.

WEIZENBAUM, JOSEPH, *Computer Power and Human Reason,* Freeman, New York, 1976.

WEINHANDL, FERDINAND (ed.), *Gestalthaftes Sehen,* Wissenschaftliche Buchgesellschaft, Darmstadt, 1960.

WHITEHEAD, ALFRED NORTH, *Process and Reality,* Macmillan, New York, 1929.

WIENER, NORBERT, *Cybernetics,* MIT Press, Cambridge, Mass., 1948; reprinted 1961.

WIENER, NORBERT, *The Human Use of Human Beings,* Houghton Mifflin, New York, 1950.

WINDELBAND, WILHELM, *A History of Philosophy,* Macmillan, New York, 1901.

WINOGRAD, TERRY, and FERNANDO FLORES, *Understanding Computers and Cognition,* Addison-Wesley, New York, 1991.

YOVITS, MARSHALL C., and SCOTT CAMERON (eds.), *Self-Organizing Systems,* Pergamon, New York, 1959.

YOVITS, MARSHALL C., GEORGE JACOBI, and GORDON GOLDSTEIN (eds.), *Self-Organizing Systems,* Spartan Books, 1962.

Index

ABOUT THE AUTHOR

Fritjof Capra received his Ph.D. in theoretical physics from the University of Vienna and has done research in high-energy physics at several European and American universities. Capra has written and lectured extensively about the philosophical implications of modern science and is the author of *The Tao of Physics, The Turning Point,* and *Uncommon Wisdom.* Currently Director of the Center for Ecoliteracy in Berkeley, California, he lives in Berkeley with his wife and daughter.